Support the Business

Delight the Fans

A Practical Guide for Sports and Entertainment Wi-Fi Deployment

By James Licata

Support the Business Delight the Fans

Disclaimer

The comments and opinions here are my own and do not necessarily reflect the views of any past or present employer.

Acknowledgements

This would not have been possible without the years of support from my colleagues, mentors and managers; special thanks to William Serpe, Richard Costanza, Art Mombert, and Ray Southworth, as well as Rock, Lameo, Ron Gula, Bhargavan Vadavur, Reshmi Yandapalli and Robert Crisp.

I would also like to thank Raymond for inspiring me to put my command of this subject into a book.

Raymond Aaron - New York Times bestselling author *Chicken Soup For the Soul* series and *Double Your Income Doing What You Love.*

To My loving wife of 20 years Lisa and our remarkable daughters, Angela and Nicola., as they carry the enjoyment of sports and technology into the future

Foreword

Enabling high capacity Wi-Fi in stadiums enhances the fan experience in ways you probably can't even imagine. Will Wi-Fi make it more likely that a fan will get off the couch and come back to the park? Absolutely!

Support the Business, Delight the Fans sketches out the technology and business issues of deploying Wi-Fi in sports and entertainment settings. This straightforward and practical guide will help you understand how enhancing the fan experience via in-venue high capacity Wi-Fi can improve all your business functions greatly and generate significantly more income.

James Licata's clarity and command of the subject ensures that you will come away with a solid grounding in the basics you need to make the informed choices for both front and back-of-the house deployment.

Making good Wi-Fi choices will position you as a revenue enabler and the principles in this book will help you drive the Wi-Fi technology value proposition to its full potential. Your IT organizational posture will increase based on your good choices at each layer of the Wi-Fi deployment, and you will deserve to be recognized as an insightful contributor to business initiatives.

Raymond Aaron

NY Times Best Selling Author

www.aaron.com

Terminology

Event

When I use the word event, I am referring to a baseball game, football game, hockey game, basketball game, soccer match, cricket match, race, prize fight, concert, theater show or any other sports or entertainment event.

Venue

A venue is the gathering place for the above-defined event. It is where like minded people gather for the purpose of being entertained, whether in a stadium, arena, amphitheater, theater, concert hall, outdoor festival, motor racetrack, horse racetrack, waterway, ballroom, convention center or any other place where events occur.

Fan

When I refer to a fan or fans, I mean an attendee(s), guest(s), patron(s) or spectator(s).

Delight [1]de·light*noun* \di-ˈlīt, dē-\

A strong feeling of happiness: great pleasure or satisfaction

Something that makes you very happy: something that gives you great pleasure or satisfaction

Introduction

Throughout my 20 plus years in information technology, I've found that the only thing that remains constant is change. We have all lived through many series of dazzling technological innovations, each one setting a higher standard for the next wave.

But none of the great technologies of our century, including the automobile, jet engine, radio, television or even electricity, has ever moved from announcement to global adoption as quickly or as broadly as wireless networking (Wi-Fi). Even within the computer technology field, the adoption rate for wireless was dramatically faster than it was for other innovations.

Deploying high capacity Wi-Fi networks in large public venues is part science and part art. It is my goal to share the technical and business knowledge I have gained during a career marked by these rapid changes and many large-scale Wi-Fi deployments. By sharing what works and truly excites the fans, I hope to guide other IT professionals in deploying robust Wi-Fi networks that both support the needs of their businesses and create a "Wow" social media experience that delights their fans.

I would like to hear your thoughts on this ever-changing field. Your feedback is much appreciated. I hope you enjoy *Support the Business, Delight the Fans*. Please feel free to contact me with comments and questions.

James Licata
Solution Architect, Sports and Entertainment Practice
Jlicata@supportthebusinessdelightthefans.com

Contents

Chapter 1
Pervasive Wi-Fi - Now an Entitlement

"Only those who risk going too far can possibly find out how far they can go."
— Emily Greene Balch

The anytime anywhere Internet experience has gone beyond an expectation - it has become an entitlement. If you doubt this, just watch a typical American under the age of 20 for 20 minutes. If they do not touch a wireless network within 15 minutes, they are probably asleep. Several waves of escalating innovation have brought us to this point of unstoppable change, which renowned author Malcolm Gladwell calls the "tipping point." With this in mind, it is evident that there is an immediate need for a pervasive high capacity in venue Wi-Fi networks.

The pace of information technology renewal was historically IT department driven. A clear path was set by equipment depreciation cycles and a department's drive for productivity. Today, the rate of change is dependent on the action of end users, i.e. the *fans*. This paradigm shift to customer driven change has resulted in Wi-Fi being adopted faster and more pervasively than any other innovation cycle before it.

How Did We Get Here, How Fast Are We Moving?

In the past, change was gradual; most people eased their way into a new technology. Recent history illustrates that the speed of technological change has continually escalated to the point where almost immediate adoption is normal. Today, technology changes within weeks, sometimes even within days! Let's take the computer for example.

The computer was developed as a byproduct of our 'race' to put a man on the moon in the 1960s. In the late '70s, computers were very large machines, encased in separate air-conditioned, glass-walled, specially built computer rooms. They used iron-cored donuts as memory bits. Only top-level scientists could access these "miracle machines" at a time until they learned to "share."

Time-sharing mainframe computers used a typewriter-like device which allowed you to enter a few instructions, then "run" your program. For me, this wave of technology arrived as I was ending high school. That is when I was first given the choice of learning to program a computer or taking a high school math class. Computers were relatively more fun as the computer language was "BASIC." Plus, the machine saved the program I created by punching holes onto a roll of paper tape. Carrying a few yards of paper tape (for the machine to read back) in my pocket to my next session was much "cooler" than carrying around a math textbook.

When I was entering the business world, business-oriented computer languages were being deployed. So, I became a Common Business Oriented Language Programmer. Does anyone else fondly remember COBOL? Or the days of mainframe computers? How about mini-computers, like the VAX? It only took up half of a computer room. They were called distributed computers and were connected like departments within a company. Electronic Mail was introduced, and people could send messages and reports by typing characters into a terminal and have them appear almost instantly anywhere there was another terminal.

By the 1990s, the world was discovering interactive computing, the Personal Computer and the local area network (LAN). These marvelous new devices started replacing fixed function terminals and distributed mini- computers. Computing started becoming 'graphical' as Windows and the MAC made computing 'personal.' Just as no person is an island, it now proved true that no computer was a computer unto itself. By that I mean we discovered that personal computers needed to be connected to each other in order to be effective business tools, so connectivity between computers was established as well. The standard became the Ethernet LAN, which was a shared media network running at a blazing speed of 10 megabits per second. This is just about when I started looking at innovation though a network frame of reference. I became a "network guy."

The TCP/IP communication protocol emerged as the winner of the unified local area and wide area protocol wars, providing us the standard for communicating between physical modalities, fiber, copper and wide area communication links. (These standards set the stage for the Internet World Wide Web wave of change, and then to Wi-Fi.) Being connected then meant plugging a wire from your device into the 'network.' The network wall plate dominated the office connection method and, by 1997, the first one megabit per second and two megabit wireless networks were being created.

The Wi-Fi we know today emerged around 2000 from this era's IEEE 802.11b standard for 11 megabit per second wireless networking.

LAN technology moved from 10 megabit Ethernet around hosts to shared multi-station Ethernet hubs to high-density Ethernet switches. The speed of these wired devices increased from 10 megs to 100 megs to one gigabit per second.

An important game-changing inflexion point occurred when end user computing devices started driving network upgrades. It was one of the first times the pace of networking change was predicated on the capabilities of the end station, not the needs of the network planners. Specifically, new PCs came equipped with 100 megabit network interface cards. Since the capabilities were not in the end stations, the user community pressured the IT department to upgrade their network to 100 megs switching.

At this point, PCs with high-speed local area network connections (LAN) and graphical interface created fertile ground for the birth of the next rapid revolution: the introduction of the Internet browser. The build out of high-speed interconnected TCP/IP networks, in combination with the standard graphical Internet browsers, connected to a large quantity of LAN sites through Wide Area Networks (WAN) — creating a web of interconnected computers and content. The Internet and this World Wide Web rapidly became the standard method of delivering data to end users (consumers). High-speed broadband Local Area Networks (LANS) became the "last mile," the edge of the Internet at work. At home there was still a place for telephone dial up modems.

The network became a common utility, like the telephone, electric lights or indoor plumbing. Companies discovered that the use of electronic content was the way to differentiate their businesses, so the corporate user base expanded rapidly. However, the stations were still tethered to the wired network. It became clear that the next major change would be a productivity-based shift to wireless mobility.

The adoption of wireless technology was modest until 2002, when network speeds jumped up to 54 megabits per second. At this point, products supporting the IEEE 802.11g communication standard emerged. The technology for mobility was present, the need for mobility was just evolving and the industry was getting ready for the adoption 'explosion' to start.

By 2005, cellular phone use was commonplace and smart phones were being introduced. Large-scale Wi-Fi deployments started inside organizations that especially benefited from having mobility capacities. Soon, Wi-Fi was in almost every college, high school and hospital. Organizations deploying voice networks over IP encouraged the development and use of the largest networks. Wi-Fi adoption was growing like a wildfire.

As people embraced a web-based social presence, the demand for network bandwidth expanded to include individual users. Network based applications became part of a user's personal life, and the next wave of change was in full bloom. Now, a large variety of web-based applications are available, with some of the most pervasive being Google, Facebook, Twitter and Instagram.

Although it appeared to have happened overnight, the true wireless explosion was precipitated by converging event. In 2010 Apple introduced hugely successful web enabled devices: the iPod, iPhone and iPad. They were designed for social interaction over the web, and specifically designed to work without a wired network connection port. These devices could only use a cellular or Wi-Fi network. There was no place to plug them in.

As these 'i' devices started selling in the millions of units, they drove an unprecedented demand for networking. And, just as there was an emerging demand to saturate the wireless Internet edge (the third and smallest of the three classes of network devices), the companies that manufacture such networking products agreed on a new wireless standard that provided a six-fold increase in wireless speed, from 54 mbps to over 300 mbps. This perfect storm of end station availability, social need for engagement and networking and capability is driving the fastest technology deployment I have ever seen.

In 2014 the demand for social media content and video games on smart portable devices deployed over high-speed Wi-Fi networks is an integral part of people's lives. It has reached the point where the current generation of fans does not look at Wi-Fi as a utility; they look at it as an entitlement. Allow me to share this story as an illustration.

Recently, while in the dentist's chair with a mouth stuffed full of cotton, I was asked what I did for a living. As I temporarily pulled one tube out of my mouth, I started telling the dentist about deploying Wi-Fi in a stadium. The dentist placed the drain back in my mouth and told me no one would want to have such a thing. It is all about the game. Without being prompted his young assistant immediately interjected, saying the game was only a part of the experience, and that connecting with her friends during games was an important part of the event. "We just have to have Wi-Fi in the stadium or else we won't go," she said. I did not even have to take the suction tube out of my mouth as she continued to make the point about the importance of being socially connected. By the end of the exchange, the dentist had conceded that Wi-Fi was important. I didn't have to say another word.

Both the dentist and hygienist were sports fans. While one may enjoy the game from the luxury box while the other may sit in the bleachers, and their expectations may vary, both will be enjoying richer experience as Wi-Fi continues to support the business and delights the fans.

Chapter 2

Design for the Business First

"Put your own oxygen mask on first."

— Airline employee giving emergency instructions

The primary reason anyone truly cares about Wi-Fi relates to its ability to drive business value; everything else is secondary. Wi-Fi is the tool a fan uses to 'touch' the brand. It also projects the organization's systems, processes and methods to an increasingly mobile workforce, thereby improving productivity and enabling 'human capital' to better manage time. In turn, this reduces costs or increases top line income, providing a positive Return on Investment (ROI) for employing the Wi-Fi network.

It is critical to recognize the two parallel needs of a sports or entertainment venue. Front of the house activities directly support the players and enhance the fan experience, with social media being an important component of the overall entertainment experience. Back of the house activities support the running of the business operation itself and are, therefore, of paramount importance. Because of this, the primary objective for Wi-Fi deployment must be business-focused, intent on driving revenue and lowering costs. After all, it makes little sense to enhance the entertainment experience if doing so isn't profitable.

Producing a high quality sports event takes coordination from many organizations; to name a few: sports operations, grounds, crews, referees, coaches, players, catering, public safety, accounting, marketing, access control, customer relationship management, the TV media, radio, working press, photographers and human resources. (OK, I admit that this is not just a few. This is a lot, and yet, still not a list of all the organizations involved.) With all these simultaneously moving parts, it is the Wi-Fi network that allows mobile communications and interactions among them. It is the Wi-Fi network that creates a foundation for increased productivity in daily activities across many departments. Email, calendaring, phone systems and database applications are among the back of house business applications that benefit from a Wi-Fi network. Thus, a Wi-Fi network must be viewed as a productivity enhancing investment, not a cost. In addition to enhancing the fan experience, the financial outlay involved in deploying the network yields cost savings and additional revenue. Wi-Fi capabilities are also instrumental in meeting today's needs and expectations of human capital involved in the business of operating a venue. The drive for productivity encourages and allows employees to use their personal devices, like iPhones, iPads and laptops for work. (The computing industry calls this Bring Your Own Device or BYOD).

In addition, interns, contractors, vendors and corporate guests who participate in coordinating venue activities often need to access secure internal systems. This is a new challenge for sports and entertainment facilities, where, historically, most computing devices were venue owned. It is a challenge that Wi-Fi networking easily solves.

To support your business successfully, first prioritize the business functions that can be enhanced using mobile resources. Determine how departments can free people from their desks to better use their time interacting more directly with customers or peers. Consider also these revenue-generating capabilities of Wi-Fi:

- High quality Wi-Fi in the luxury suites raises the level of concierge service, better supporting corporate functions and rental for non-game presentation events.

- Mobile concessions can generate additional sales and impulse purchasing by placing the product more directly in front of customers.

- Flexible access control, moveable turnstiles and gate ticket scanners support faster fan entry into the venue. This increased time in the venue can translate into more time shopping in the team store and increased food and beverage consumption.

- Food ordering applications can help fans bypass waiting lines and allow more fans to visit the point of sale register in any given amount of time.

Wi-Fi deployment must be able to address the needs associated with getting employees and guests up and running on the network; this is also known as on-boarding. Because of the traditionally small size of a venue IT staff, on-boarding needs to take place with minimum IT direct participation.

Get a good Wi-Fi on-boarding program in place. Once your IT staff is on-board they can help the other business functions to do the same.

Supporting the business usually produces direct measured return on investment. Let us now explore some of the ways to monetize the indirect revenue sources. This involves creating a "wow" network enhanced social experience for the fans.

Chapter 3

Social Media and the Fan Experience

"Take me out to the ballgame; take me out with the crowd;
Buy me some peanuts and Cracker Jack; I don't care if I never get back".
— Jack Norworth and Albert Von Tilzer

Upbeat attendees in the stadium can attract even more ticket buyers to the venue, especially if their level of excitement can be communicated via the Internet. Crowd interaction helps achieve an enhanced personal connection between the attending fan and the event. The Internet creates a similar connection between the event, the attendees and all the other fans. It is easy to see how venue business executives appreciate the monetary value of an enthusiastic, happy crowd.

Creating an atmosphere is not only about generating as much noise as possible in the stadium; it is also about the volume of online buzz, especially within the social media sphere. It is also about making the fans feel they are part of an event, giving them a multi-sensory communal experience they could not get by watching it on television at home

Like Minded People in the Moment

Fans in the venue project their experience through many actions, such as cheering, standing, banner waving and horn blowing. The way a lively crowd boosts the home team and intimidates the opposition can also help bring success and money to a team. This is one reason venue designers put so much effort into creating the seating bowl, which allows the vast majority of people in the crowd to feel close to the action. In fact, the designers' goal is to make even the last row feel connected.

Today's crowd is also devoted to sharing the moment electronically. Fans love to take pictures and post them on social media during a game using

Internet applications including SMS messaging (texting), Twitter, Facebook, Foursquare, Instagram, Flicker, Tumbler, Pinterest and email. Plus, there is a plethora of new applications emerging for 'sharing the moment.' Electronic buzz is as much about sharing the joy or frustration and the emotions of the moment, as well as the facts.

Creating Content is Staying Connected

Through the use of social media applications and the Internet, individuals are able to create their own content and comment on any issue in the world of sports at any time. Perhaps the most important effect of these technologies, however, has been greater audience empowerment. Individuals use technologies to become actual participants in the sports communication process.

It could be said that social media has expanded the 'engagement' of the sports conversation beyond the game itself. People talk about teams, players, and organizations 24 hours a day, seven days a week. Sharing the game experience starts well beforehand when a fan lets their friends know they are going to the game. The expectation is set that they will share this event with the ones who cannot go. The intensity of this conversation peaks when fans are in the venue during the game.

When your team wins, you are happy. Posts on Facebook expressing that happiness can easily lead to fresh conversations between friends. Twitter allows fans to interact with complete strangers and discuss their favorite team, a particular game or a specific player. This provides a nice change from talking to the same people about the same thing.

When a team loses, a player leaves the team or gets hurt, or something else bad happens social media gives fans a place to vent. This exchange of viewpoints allows fans to release their frustrations and be heard, all while increasing their feeling of being part of the experience.

Twitter and Facebook pages have also opened up a new, more direct line of communication between fans and the team. Many sports fans have tweeted to their favorite team, player or sportscaster/radio host. While many fans will never get to talk directly to their favorite athlete, being able to tweet them actually feels like a way to do just that!

Whether it is Facebook, Twitter, Instagram, Pinterest or any other social media platform, the ability to interact with new people is part of what makes social media so popular. Being able to use a hashtag (the symbol # at the front of a word or phrase that labels your tweet or post so anyone interested in the subject can easily find and read what you have to say) to join a conversation where most people are fans of your team is very entertaining, and can lead to more followers or Facebook friends.

Staying connected with social media via smart phone is also a way to keep current, and sometimes create, "breaking news," i.e. anything of interest to other fans. Some information is disseminated via Twitter well before it is

heard over the radio or seen on TV. At times, Twitter breaks sports story well before ESPN.

Bill Rasmussen, (left) Founder of ESPN and the 24-hour a day sports conversation, is a visionary.

Years ago Bill realized the importance of the sports dialog. There is now the ESPN app that allows fans to interact instantaneously with the sports news network anytime they want, anywhere they want.

People learn a lot of breaking sports news, such as player and personnel transactions, injuries and game updates this way. It is the news medium of choice for many younger fans, especially as it allows them to stay up-to-date while mobile and not near conventional media like radio or TV.

Fans definitely do not want to miss any of the action, and love seeing the game from multiple camera angles. This opens the door for creating and providing league or venue specific smart phone applications. When every handheld device is also a handheld TV, the expansion of in-venue video content is soon to be a major factor in Wi-Fi needs.

The handheld TV has been referred to as the 'The Third Screen' and, at no time has this been truer than today, when smart phones, iPads and other electronic gadgets permeate our every experience and blur the lines between virtual and actual reality. Are you going to believe what you see on the field or the alternate angle replay you see on your smart phone?

The Third Screen - Picture by - Aruba

As social media continues to evolve, it is quickly creating better, easier technology for users, especially with respect to video. (And, with video comes the necessary Wi-Fi bandwidth to support it.) Twitter's latest foray into video is their new app, Vine, which allows users to record six seconds of video and share it to a news feed, essentially broadcasting video-tweets without needing to load an attachment. Consider how Snapchat, the iOS and Android photo-sharing app used by millions to send self-destructing media files can be used. These new apps have the potential to compete quickly with the likes of Instagram, and to do so in the same way that Instagram made Facebook nervous enough to acquire Instagram for an above market value.

Cultivate Interaction, Then Gather Responses

As of 2012, Twitter has over 100 million active users worldwide, generating close to 230 million tweets per day. Encouraged by sports shows that now make social media a major component of their programming strategies, fans and athletes have proven to be some of the most prolific and adept users of Twitter and other social media platforms. They thrive on having their posts and messages shared with the world.

The concept of gathering fan social media responses is similar to having fans vote for their preferences in more traditional media, except that it all happens in real time. The votes are not counted as in a ballot but are measured by volume, including the number of tweets, Instagram picture posts, video, text messages or emails.

Some stadiums are already taking steps to optimize for social media. Manchester City in the UK is enabling Etihad Stadium for social media interaction within the actual game. A photo tweeted from the official @MCFC account shows what appears to be a TV screen in the stadium concourse displaying tweets with the #BlueView hashtag. Presumably, fans can use the hashtag for a chance to have their tweets displayed, in this case throughout the stadium.

Outside the US, there is a buzz about electronic in-venue wagering, which is becoming a normal method of fan interaction and revenue generation. This application of robust fan facing Wi-Fi facilitates a more exciting personal experience for fans by reducing their fear of placing an incorrect or out of time wager because they left their seat to place a bet at a kiosk or "window." The immediate ability to bet also increases wager volume and supports venue profitability.

An Important Review

Digital media has created and facilitated new channels for sports fan engagement, as well as for enhancing the fan's perception of participating in America's sports culture. High profile athletes are also using social media to increase their fan base and promote their own celebrity status, thereby creating the sense that they are more accessible to their fans.

The third screen continues to permeate our experiences, blurring the line between virtual and actual reality. As the technology advances, teams positioned to harness the opportunity by upgrading their stadium wireless networks for inbound and outbound video distribution will grow their brands while those who ignore this technological upgrade will struggle to participate in this expanded engagement model.

Human emotions tend to develop in similar phases around new Wi-Fi deployment developments which, to paraphrase Paul Gainham of Juniper Media, are:

- **Phase 1 Excitement** — the service is new, cool and the latest 'must have.'

- **Phase 2 Expectation** — the service is widely available; it is mainstream.

- **Phase 3 Exceed Expectations** — I expect and want to see more.

The lesson for Wi-Fi enabled venues is that while capacity and coverage are indeed vital they are not necessarily valued on their own. The key to long-term value differentiation lies in the delivery of applications beyond the basics.

IT leaders, this is a call to action to engage with the social media people in your organization, as well as those in the customer relationship management (CRM) department. Join together in the deployment of a robust and pervasive Wi-Fi network in your venue.

My Commitment

As the technology advances, it is my belief that teams positioned to harness the opportunity by upgrading their stadium wireless networks will grow their brands while those who fail to do so will miss out on a significant business growth opportunity. Besides being able to help with the design, I can also help bridge any gaps in communication between businesspeople and technologists. As Wi-Fi deployments expand it is my goal and my passion, to have no venue left behind.

For the latest full color edition of this book and its supplements, reach out to me at www.supportthebusinessdelightthefans.com.

Chapter 4

Engage from Street to Seat

"The handshake of the host affects the taste of the roast"
— Ben Franklin

When does the stadium experience begin? It begins when a ticket is purchased, and anticipation starts to build. As soon as a fan tells the first family member or friend that they are going, a social experience is created.

Make It Worth the Cost

The fans know there is a real cost to attending an event, both in time and money. They are willing to spend a significant amount of both to be part of the live experience. The Team Marketing Report[1] breaks out the average cost for a family of four by various sports and markets. According to the site's Fan Cost Index (FCI), going to a major league baseball game can cost up to $339, while attending an NFL football game is around $617. These amounts are not insignificant to the average fan, and prices continue to trend upward.

---

[1] (*https://www.teammarketing.com/btSubscriptions/fancostindex/index*)

Team Marketing Report • April 2013

team marketing report

Team	Avg. Ticket	Pct. Change	Avg. Premium Ticket	Beer[1]	Soft Drink[1]	Hot Dog[1]	Parking	Program	Cap	FCI	Pct. Change
Boston Red Sox	$53.38	0.0%	$172.51	$7.25	$4.00	$4.50	$27.00	$5.00	$18.99	$336.99	0.0%
New York Yankees	51.55	0.0%	305.11	6.00	3.00	3.00	35.00	5.00	25.00	324.30	0.0%
Chicago Cubs	44.55	-3.9%	106.98	7.25	3.50	4.50	25.00	5.00	20.00	298.20	-0.7%
Philadelphia Phillies	37.42	0.0%	79.82	7.75	4.00	3.75	15.00	5.00	17.99	257.16	0.0%
San Francisco Giants	30.09	10.6%	86.63	6.75	4.50	5.00	20.00	5.00	18.00	237.87	7.8%
Washington Nationals	35.24	15.4%	192.89	8.25	4.75	5.00	10.00	0.00	15.00	236.46	19.3%
St. Louis Cardinals	33.11	3.7%	77.26	6.75	5.25	4.25	10.00	2.50	16.00	230.94	2.7%
Miami Marlins	29.27	-4.0%	116.18	8.00	4.50	6.00	15.00	0.00	19.99	230.05	-5.1%
Toronto Blue Jays	25.38	0.1%	70.06	7.34	5.07	5.57	15.70	5.07	22.89	229.84	3.6%
Houston Astros	30.09	0.0%	51.24	5.00	4.50	4.75	15.00	4.00	16.99	224.33	0.0%
New York Mets	25.30	-7.1%	83.78	5.75	5.00	6.25	20.00	5.00	18.00	223.70	-2.9%
Minnesota Twins	32.59	-1.4%	74.18	7.50	4.00	4.00	6.00	3.00	15.00	221.36	1.7%
Chicago White Sox	26.05	-10.2%	86.94	6.50	4.50	3.75	20.00	4.00	15.99	210.19	-5.7%
MLB LEAGUE AVERAGE	27.48	1.8%	90.48	6.09	3.67	4.14	14.06	2.99	17.21	207.80	0.6%

If you'd like a copy of the full Fan Cost Index report, please visit [1]
(*https://www.teammarketing.com/btSubscriptions/fancostindex/index*)

The fan's expectation, obviously, is that going to the event will be a much better experience than staying home. It is the venue's goal to provide a *great* experience, one that lives up to the fan's expectations. To deliver this, organizations must strive to be the best on the field and in the stands.

Everyone Wants to Stay Connected

People's need to stay connected everywhere is evidenced all around us. Have you ever been driving and seen another driver on the phone or texting? The act is so ubiquitous that the better question might be when haven't you seen a driver on the phone. And, when was the last time you heard a parent tell their child to leave the cell phone off? Media and mobility drives the desire to stay connected and to share the moment's experience, whatever it is. Connectivity, specifically cell phone coverage, is not something people want to be without.

Fans may not currently expect to have good cell or data coverage in your venue, but they want it. And, they expect it to get better over time. The natural tendency will be for fans to compare their in-stadium experience in one venue versus connect-ability in other venues, as well as to their home or office connections. In other words, fans believe "my phone, tablet, or any other device I bring to the stadium should work well."

The fans' expectation (as the technology gets more pervasive) creates a race between venues to expand the Wi-Fi experience and to provide better service. Fans also stay at hotels, travel through airports and go to schools that all have increasingly high quality, free wireless coverage. They see being connected while mobile as the norm. Wearable computers such as the Google Glass will only accelerate this expectation.

Staying socially connected also drives the need for access to wireless data. But where this is not yet possible, a stadium must at least provide cellular service. You cannot depend on cell carriers to make this happen for you. Wi-Fi can provide both phone and data accessibility. An excellent example of how the two are used in tandem is an online banking application that uses a secondary verification process. In that type of situation, you might input data, pay bills or perform another action that requires a second password or code. For security reasons, your bank uses cellular text messaging to deliver that code to you.

The Verizon Center - Washington D.C.

It is not where we should provide good cellular or good Wi-Fi coverage, but how can we provide both and use them together to enhance the interactive social Game Day experience.

The Fan Portal: Revenue Enabled

Think of the fan portal as a form of digital on-boarding or digital-seating. Like a players' reward program at a casino, but without the inconvenience of a key card, the portal would provide a welcome screen for fans to sign-in with their email addresses and seat numbers. The portal, like an excellent announcer, lets the fan know an excellent experience is about to begin.

Bruce Buffer —(right) the voice that gets you started - Let's Get Ready to Rumble UFC - Michael announces similarly for HBO Boxing

The fans can immediately begin interacting with other attendees, download the starting roster and players' stats and interact with the venue's communication system. Hash-tagged tweets could be displayed on the Jumbotron as could personal welcome messages, etc. At its most developed point, the portal would allow fans to place orders for food, souvenirs, etc. to be delivered to their seats.

Like the ticket taker at the gate, a website welcome page can be either cold and 'all business,' or warm, helpful and friendly. Venues successful at creating a great fan experience allow the live greeter to provide information about the team and guide attendees towards their seats or a rest room. If you choose to use a fan portal welcome screen, it should follow the same line of thinking. The presentation of a gateway web page (the Welcome screen into the fan portal) should make fans want to connect and encourage them to enter the park electronically and get seated.

A 'fan portal' allowing attendees' free access to your venue's Wi-Fi system can enhance their experience by providing what some might term a type of concierge or VIP service. For the fans, it recognizes them as individuals and makes a positive first impression, welcoming them to the event and encouraging their interaction with the team and each other. For the venue, it enables opportunities for generating additional revenue through customized food and souvenir offerings at seat-level (which, by the way, further enhances the fan experience).

Financial Benefits of a Loyalty Program

There is a school of thinking that says fans should pay a nominal fee for Wi-Fi network access, with pricing similar to that at some hotels and airports. This does help recoup some of the cost of the network's deployment, but you can achieve more long-term revenue from a fan loyalty program enabled through your Wi-Fi than from a pay-per-use approach. A loyalty program would provide an incentive for fans to connect and the financial advantages associated with their being connected could be substantial, even in the mid-term. So, rather than charging, consider implementing a point system that rewards fans with a free promotional item or refreshment. Instead of collecting credit card information to feed a billing system, ask for a ticket number and email address to feed your fan analytics engine. The conversion of a fan from unknown association to a valid identity will do more to enhance the lifetime financial value of the fan relationship than the single sale of Wi-Fi service.

In terms of fan perception, a loyalty welcome page also creates a different, warmer greeting. Consider how you feel when asked to pay for Wi-Fi at a hotel, airport or conference. How many times have you or someone you know chosen a coffee shop or other public venue with free Wi-Fi access rather than pay the service charge elsewhere? While the pay-per-access model may still be viable in specific situations, the world is already well into the 'free for information' model.

Wi-Fi and Cell Phone Interaction

The fan portal can be a place where cell phone and Wi-Fi network complement each other for your marketing and information gathering purposes. Cell carriers know the cell phone numbers of all stations connecting in or around your stadium; they also know the name and billing address associated with each of those phone numbers, as well as the Wi-Fi hardware address of each cell phone. However, the carriers will not or cannot provide this data to venue owners who, until now, have been at a disadvantage in terms of real-time marketing opportunities.

One of the goals of a venue owned Wi-Fi network would, then, be to convert an unknown wireless association to a known fan (with the fan's permission, of course). One method of accomplishing this would be to ask the fan for a valid email address or phone number to which a Wi-Fi password can be sent, thus establishing a valid in-venue identity that can be used for data mining.

Data Mining

Greater and more accurate customized mapping can be employed for revenue generation and greater efficiency in venue management. Once the Wi-Fi environment begins feeding useful data to the fan analytics department, the network becomes an extension of the marketing and sales force departments. The information gathered can be used to encourage refreshment and souvenir purchases as well as fan loyalty to the in-venue experience.

The information gathered has significant additional value when viewed through 'big data' applications. The usefulness of mined data extends to arrival time analysis, the tracking of fan Internet usage habits and better time and space management. Fan arrival times can be historically aggregated for sharing with mass transit or municipal public safety organizations to improve ridership patterns and equipment or manpower allocations. Similarly, a venue can use this data to refine work hours and re-allocate staff location based on actual need.

Additionally, tracking and aggregating fan online movements allows for better interaction with new and existing sponsors. For example, if a large number of fans go to a specific sports or energy drink site during the game, that company can be contacted about sponsorship, as can a competitor within the same product class.

CRM - Lifetime value of a fan

A key goal for venues is to generate repeat attendance and increase sales per fan. The financial worth of that fan over time and his or her behavior is sometimes referred to as the lifetime value of the fan relationship. Data that helps the organization fine tune its offerings based on fan preferences can prompt faster, larger and more frequent sales, thereby increasing the fan's lifetime value to the venue.

In addition to the use of attendance data, fans' purchase history can be used to refine the promotional techniques, give-away items and rewards employed by long-term loyalty programs. As venues provide game-specific promotional offers, they want to ensure their product choices represent the best chance for loyalty program success and sell-out of items. Knowing what repeat attendees are looking for provides CRM managers the opportunity to hone in more efficiently and effectively than they could with a random canvassing approach. In the case of direct touch sales, having a more accurate knowledge base can increase productivity of the offer presentation process.

Analyzing and aggregating smaller pieces of information can lead to bigger conclusions and larger revenue opportunities. Having an ongoing feed of valid presence and purchase data can open cross marketing opportunities for affiliates and merchandizing. Identifying a group of fans that regularly attend events in a venue, and understanding their behavior pre and post-game (or concert), can lead to cross-promotions with area restaurants or other types of events held within the same venue and off-season activities.

As you can see, collection of fan behavioral and purchase data provides an enhanced fan service and enables multiple revenue streams for the venue.

Chapter 5

Design for Capacity Not

Just Coverage

"I have five bars, and I still can't connect!"
—A disgruntled wireless user

Pervasive coverage for clubs, concession areas, service areas, team rooms and working areas of your venue is essential. But, having Wi-Fi everywhere in your venue is only the beginning of the enhanced fan, player and staff experience. Coverage alone is insufficient in the new Wi-Fi paradigm. Driving high signal levels through a powerful antenna does not ensure that a handheld device will work clearly or responsively.

When coverage is available, but performance of Wi-Fi is iffy because too many people are accessing the network, you end up with a situation in which there is asymmetrical communication. That means one may receive information well, but not be able to respond without a dropped connection, slow load time or some other unacceptable performance issue.

Design in Symmetry and Balance

The situation is similar to being on a boat at night, with just a flashlight for communication with shore. You may be able to see the light from a nearby lighthouse because of the size and power of its beam, but people in the lighthouse may not be able to see the glow of your little flashlight. As a result, your experience in the boat is not satisfactory. Likewise, Wi-Fi users must receive a strong, clear, actionable signal no matter how demanding the network's usage load.

Think Capacity

A few years ago providing coverage was the sole goal of establishing a Wi-Fi network. Today the primary network design consideration is wireless station capacity. It's critical to provide a sufficient amount of bandwidth to all areas for every point in time. Usually this requires multiple layers or blankets of Wi-Fi coverage to handle the over the air channel capacity required for good service.

Channel Planning Photo from Aruba Networks

In addition to determining the number of wireless stations and users in an area, planners must take into account the reasons for usage when calculating capacity needs. These capacity calculations can first be made based on the estimated number of total users in an area, and then refined by the type of traffic being generated. However, you would be wise to consider starting the network design in your venue by defining the primary usage for each area including the footpath, food service, clubs, function rooms, media areas, corporate and administrative offices, ticketing and the general seating areas.

For large areas, such as the seating bowl, loads have to be allocated per section to determine the capacity for the area. The following pages lay out some of the unique requirements for specific areas.

RF Channel Capacity

One of the key fundamentals in planning out Wi-Fi implementation is being knowledgeable about which available radio frequencies (RF) you are allowed to use. RFs are considered unlicensed frequencies and are classified by regulatory authorities and the Federal Communications Commission (FCC) in the US as industrial, scientific and medical (ISM) radio bands.

Wi-Fi is restricted to two small number frequency bands of the radio spectrum, 2.4GHz and 5GHz. Communications equipment operating in these bands must be able to handle, or 'tolerate' any interference generated by ISM equipment in order to provide reliable service, especially as users have no regulatory protection from the operation of an ISM device. Devices operating on a 2.4GHz band have a smaller range of frequencies available to them than those on a 5GHz band. Plus, those using 2.4GHz can interfere with each other at greater distances than devices using 5GHz.

Currently there are more 2.4GHz smartphones and tablets being used in the general seating areas while there is usually an approximately even mix of 2.4GHz and 5GHz devices observed in the suites and press boxes. That ratio is changing as more 5GHz and dual 2.4/5 GHz devices come onto the market and, eventually, the mix is expected to shift heavily toward 5GHz devices.

With this information in mind, the IEEE designed 802.11 AC as a 5GHz standard only. (The reasoning behind this decision and the capabilities of 802.11-AC are detailed later in this chapter along with a section on available frequencies and how they are used in Wi-Fi deployments.)

2.4 GHz Channels B/G/N

A Wi-Fi station must choose a specific channel (a portion of an available frequency) on which to operate. In the 802.11 B and G specifications, the 2.4GHz spectrum is divided into 14 22MHz wide channels. Overlapping coverage is not acceptable as it causes co-channel interference (CCI) that, in turn, causes a severe degradation in performance and throughput. If overlapping coverage cells also have frequency overlap, frames will become corrupted, retransmissions will increase, and throughput will suffer. For these reasons, 25 MHz of separation is required between the center frequencies of channels.

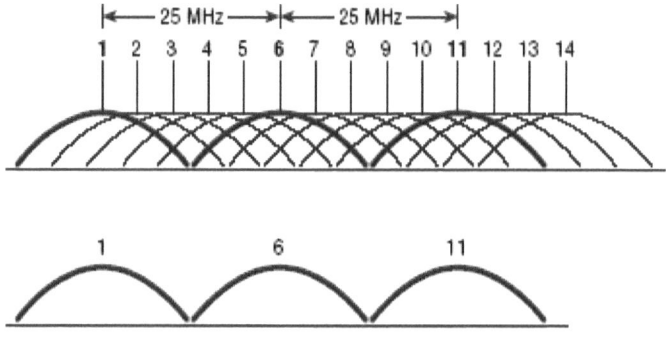

These Three Channels are Non Overlapping

Three of the most commonly used non-overlapping channels are 1, 6 and 11. Channels 2 and 7 are also non-overlapping, as are 3 and 8, 4 and 9 and 5 and 10. The important thing to remember is that there must be five channels of separation in adjacent coverage cells. For example channel 1 and 6.

Some countries use all 14 channels in the 2.4 GHz ISM band but, due to positioning of the center frequencies, no more than three channels can be used because of the need to avoid frequency overlap.

5GHz Channels A/N

5GHz band channels are only 20Mhz wide. As defined by the IEEE, there are currently 12 channels available, and they are divided into UNI bands. All 12 of these channels are technically considered non-overlapping because there is 20 MHz of separation between the center frequencies. There will be some frequency overlap of the sidebands of each channel, but all 12 channels can be used in a channel reuse pattern.

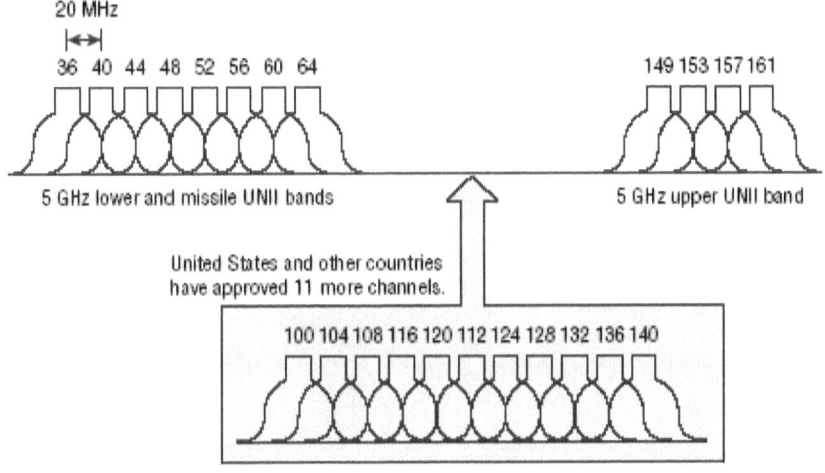

23 Total Channels are available in 5Ghz

Channel Reuse — Layers or Cells

In design alternatives without a channel coordination mechanism, configuring multiple access points all on the same channel will cause significant signal degrading and loss of performance. This condition is called co-channel interference. Data corruption is caused by your own APs transmitting at the same time over shared frequency space. The end result is decreased throughput and increased latency. Adjacent cell interference is RF interference caused by your own APs (also known as co-channel interference).

To avoid CCI, a channel reuse or a coordinated channel design is necessary. Overlapping RF coverage cells are needed for roaming but, again, overlapping frequencies must be avoided. The ability to transmit on a total of 23 channels in 5GHz allows legacy interfering designs much more flexibility.

In the United States the only three channels that meet these criteria in the 2.4 GHz ISM band are channels 1, 6 and 11. Therefore, when deploying Wi-Fi in overlapping coverage cells, a three dimensional model for channel reuse patterns (similar to the one pictured below) is required. A site survey must be performed on all levels, and the access points often need to be staggered. What is a site survey? How should I effectively conduct one? This could become a chapter of its own.

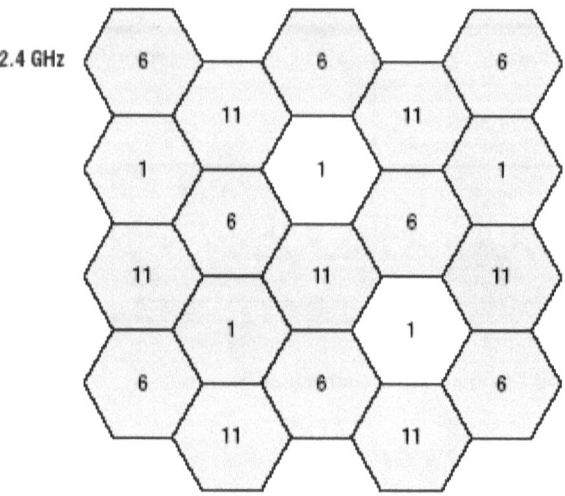

Typical 2.4 GHz Channel Plan

Channel reuse patterns should also be used in the 5GHz UNI I bands, which allow more flexibility to the larger choice of channels. Due to the frequency overlap of channel sidebands, there should always be at least two cells between access points on the same channel.

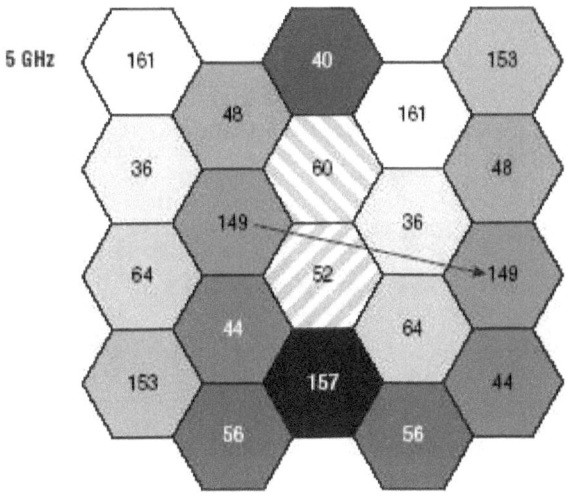

12 Channel plan- 2 channels of separation

In a high-density area such as a sports venue, many APs will, therefore, end up on the same channel. If these APs do not have a control channel for coordination, medium contention overhead occurs. Wi-Fi uses a listen-before -you-talk technology called the clear channel assessment (CCA) to ensure that only one radio can transmit on the same channel at any given time. If an AP on channel 1 is transmitting, all nearby access points and clients on the same channel will defer transmissions. The result is that throughput is adversely affected; nearby APs and clients have to wait much longer to transmit because they have to take their turn.

Adjacent channel congestion is the worst type of Wi-Fi interference. It has been determined that adjacent channel contention degrades performance more than CCI. So, co-channel congestion is preferable to adjacent channel congestion because of the way the wireless conversations are managed.

To illustrate, think about being at a concert where a band is playing loudly while there are tons of people, each talking with their own group of friends. With this much going on, it's difficult to talk to your own friends and, when you start to talk louder, the person next to you has to raise his or her voice to talk to their group. Since you're hearing multiple conversations, as well as music from the band, it seems impossible to communicate.

A visual representation of where neighboring wireless access points are active is an invaluable tool when planning your own network. It's easy to see how chaotic adjacent channel congestion is compared to co-channel!

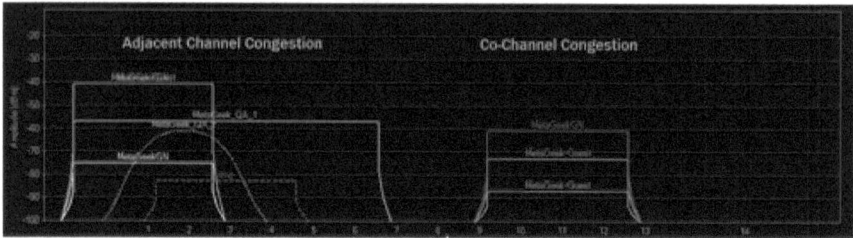

Adjacent Channel Interference - Source: metageek

Client station transmissions also result in a medium contention performance downgrade. If a client (a wireless station) is at the outer edges of a coverage cell, the client's transmission may propagate into another cell using the same channel. All of the radios in that other cell will defer if they hear the original client's transmission. It is, therefore, fairly important for the network to assist the client to connect to the best AP.

Protocol Capacity

The protocol used in wired networks and the Ethernet is governed by IEEE 802.3 standards and uses Collision Detection and Collision Recovery processes to ensure proper use of the medium. Wi-Fi has to use Collision Avoidance because it uses a shared open medium, the RF spectrum (often referred to as the Air), instead of a closed medium like copper or fiber.

Ethernet is described as CSMA/CD Carrier Sense Multiple Access/Collision Detection. Wi-Fi is described as CSMA/CA, where the CA is an acronym for Collision Avoidance. That change from Detection to Avoidance makes a big difference.

The CSMA common to both mechanisms simply indicates that sending devices look to see if the network is free or sense the information Carrier Medium before transmitting. The Multiple Access Technology implies many users are free to access the physical medium.

The Wi-Fi standard —802.11 coordinating — is determined primarily by the IEEE and defines how a station will contend for their share of the radio spectrum. With respect to Wi-Fi, contention management is a distributed coordination function. It's called 'distributed' because there is no master or central device coordinating access to the medium. Control is shared or distributed in a statistically fair way between all devices. This is intended to create "fairness" irrespective of whether devices are high-speed ultra-modern or slow legacy ones. It is very important to note that the mechanism is based on the winning of a contention round, not the allocation of time.

We therefore have to look at Collision Domains as the range of possible stations that contend for the medium.

The Wireless Clients

A station wishing to transmit first senses the channel, waits for a fixed period of time and then, if the medium is still free, transmits. After waiting a short interval, the transmitting station sends an acknowledgement frame to the receiving station. The receiving station confirms receipt of the data with an acknowledgment frame. For every successful transmission a successful acknowledgement is required. This method of either sending or receiving at any specific time is referred to as a half duplex protocol. Only sending or receiving can occur at any given time, reducing the available throughput by approximately half. While there are both protocol optimizations that improve this number and overhead mechanisms that reduce it, a good general rule is 300 megabits per data link will deliver a maximum throughput of approximately 150 megabits per second of useful data.

If two stations transmit at the same time (a collision), the access point will respond with an acknowledgement to one of the two stations. The station not receiving the acknowledgment will back off for a random amount of time and listen for a quiet moment to try again. This process of sending a transmission again is referred to as a retry, and the process of waiting and trying again is called contention. By the rules of the Wi-Fi protocol, each station, including the access point, has a statistically fair chance of contending and winning a chance to transmit. Reducing the amount of collisions and retries can, therefore, be used as a mechanism for increasing the Wi-Fi network's overall capacity.

If a collision is experienced by a transmitting station, it doubles the random number it originally generated, progressively increasing that number up to a maximum number. This can be is a very long time in the data world. Any other station sensing that the medium is busy will temporarily pause its back-off timer and not resume it until the medium is sensed to be free.

Capacity — Data Rate, Fairness and Bad Clients

One of the things the standard does not specify control of is the client station. According to the Wi-Fi protocol, all clients are free to transmit on any channel at any time, and the network cannot regulate their signal strength or its propagation. Depending on the physical environmental factors, a client's signal can be heard at up to 300 feet. A typical wireless IEEE 802.11n access point with a standard antenna might have a range of 240 feet indoors and 600 feet outdoors.

In a traditional micro-cell design, multiple independent radios are placed in close proximity to each other. Because of the CA rule, no station in either cell can transmit while it can hear the transmission (Carrier Sense) of the client in another cell on the same channel. Again, this situation is known as co-channel interference.

A station at the center of a cell (assuming the center to be the strongest signal) will be able to converse at the highest bit rate. Let's say it's a 300Mb/s capable AP and STA. When the client STA gets progressively closer to the edge of the cell a process called rate adaption takes place and the bit rate progressively drops through the available rates (300, 270, 240 ... 11,5,2,1). This impacts the entire cell as the "fairness" calculation will allow each STA an equal number of frames per period irrespective of speed. This means that an edge client could hold up the entire cell 130 times longer than a STA at the max bit rate.

The problem is multiplied in an environment where a client is very mobile and closes whichever channel and access point is associated with it. As some such clients move, they choose to hold onto their attachment well beyond the point where a better signal from a surrounding access point becomes available. In this type of situation, the station is said to be sticky to the access point. Having an environment that can identify and guide a client to the best possible AP will, therefore, perform at a higher capacity than one that merely allows stations to be sticky to the access point.

The same 802.11 'fairness' algorithm also applies to clients with older technology. An architecturally slow client like the 802.11b is located dead center of a 300Mb/s cell. It will still only achieve 11 Mb/s, so it will also slow the entire cell. Clients moving away from a cell center may choose to stick to the previously good cell way beyond its useful value until it is transmitting at low data rates. For this reason, some venues have stopped supporting lower data rates to improve the overall performance of the venues network.

BMW Calculations

Now that you have some background information and know a bit about the considerations that must be taken with regard to frequency capacity, protocol capacity and some client imposed limitations, you can start thinking about capacity for the different usages you plan to support in your venue.

Let's start with an example that assumes there will be 100 Wi-Fi users in the Press Box, all with laptops. You can determine which bandwidth will be required by analyzing three scenarios: best case, most likely and worst case.

Assume that data throughput will be in a Wi-Fi protocol environment where only half of the transmission bit rate is available for useful data. A user with 300mps connection on their laptop can expect a maximum of 150mps of data to be passed to a station on the wired network under ideal conditions. This is a rough translation of saying Wi-Fi is a half duplex protocol.

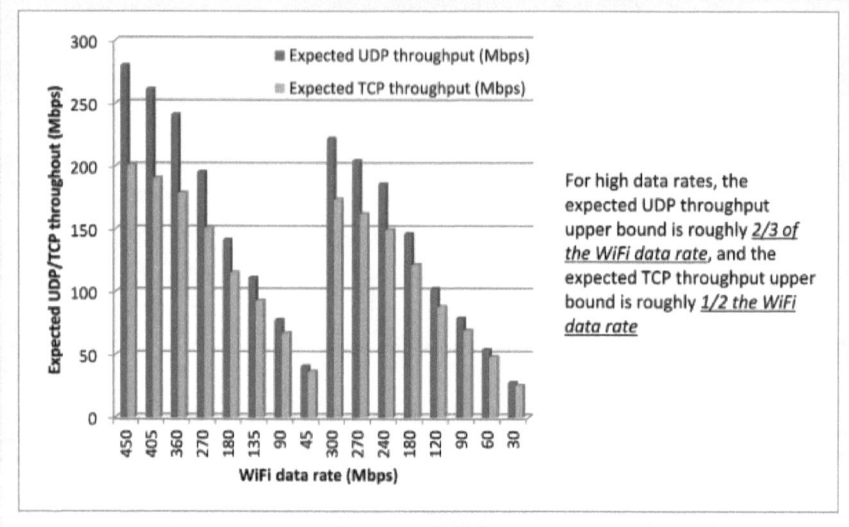

For high data rates, the expected UDP throughput upper bound is roughly *2/3 of the WiFi data rate*, and the expected TCP throughput upper bound is roughly *1/2 the WiFi data rate*

Expected Throughput

Throughput Graph - Source: Bhaskaran Raman, PhD –blog post - CWMP

Using mathematical models, Bhaskaran Raman, PhD of ITT, has determined that — for high Wi-Fi data rates such as 300 Mbps or 450 Mbps — the *expected UDP throughput upper bound is roughly two-thirds of the data rate* while the *expected TCP throughput upper bound is roughly half of the data rate*. Adjusting for further effects, such as retransmissions due to packet errors, signal strength and data rate variability, throughout upper bound can go down by another 20-30% or more.

You will also need to consider end station usage when figuring out the different scenarios. Returning to the example used earlier, in addition to reviewing the devices being used, you will also have to make assumptions about the work those 100 people in the Press Box are doing.

The table below provides Wi-Fi load guidelines for a typical laptop device listed by application type.

Application	Range of Throughput
Audio	100 – 500Kbps
Backups	10 – 50 Mbps
File Sharing	1 – 10 Mbps
Online Testing	1 – 3 Mbps
Printing	1 – 2 Mbps
Video – Standard Definition	.5 – 1 Mbps
Video – DVD Quality (480p)	2 – 3 Mbps
Video – High Definition (720p)	4 – 6 Mbps
Video – High Definition (1080p)	10 – 12 Mbps
Web	.5 – 1 Mbps

Throughput by Application, Meru Networks

For the worst case (most consuming) option, calculate the capacity of 100 stations streaming an Internet video at medium good quality resolution, 480p. The calculation would look like this: 100 users * 2megs = 200megs required. This formula indicates that one 300megs Wi-Fi channel delivering 150megs of throughput would not be sufficient to carry this workload. A second access point, far enough away for simultaneous transmission (special reuse) on the same channel, or an AP in close proximity on another channel would need to be added. If the access point was offering 150megs channels with 75megs of useful throughput, four access points would be required.

The most likely case would have 60% of the stations surfing the web, 25% streaming video and 15% engaged in local device activities such as writing. The associated mathematical formula would be: 60 * 1meg + 25 * 2megs + 15 * 0megs. The result would show that 110megs of throughput would be required in the air. This amount can be delivered on either a 300megs channel or two 75megs channels. Two access points would be required.

The best case scenario would have 50% of the people surfing the web, 10% streaming a video and 40% watching the event or engaged in local activities. The load would be 50*1meg + 10 *2megs or 70megs total. This model, which is highly unlikely, would require only one access point.

That explains why they call it the Best Case

The best, most likely and worst case bandwidth analysis shows that you could deploy four, two or one access point in our Press Box example. From this, you can conclude that two access points would be the most reasonable deployment approach. The scenario also allows you to generalize a capacity model for this type of area to be 50 laptops per access point.

When protocol overhead is added, the calculations do not remain exactly linear. The table below is very helpful in calculating the number of Wi-Fi channels required by application type to support a specific number of stations using 20Mhz wide IEEE802.11n channels. These 150mbps data rate channels carry 75 mbps of useful data.

Required Number of 20 MHz Channel Layers										
Client Throughput	Number of Clients									
	100	200	300	400	500	600	700	800	900	1000
0 - .5 Mbps	1	2	3	4	5	6	7	8	9	10
.5 - 1 Mbps	2	4	6	8	10	11	12	13	14	15
1 - 2 Mbps	4	8	12	16	20	22	24			
2 - 4 Mbps	8	16	24							
4 - 8 Mbps	16	24								

Channels Required for Client Capacity – 2012 Meru Networks

Chapter 6

Practical Considerations by Function

There will be some Wi-Fi applications that achieve critical business status because of their value in terms of increased productivity through mobile-based face-to-face interaction between service staff and fans. Here are two common examples involving the use of handheld scanners: 1) for ticket verification (often referred to as access control), and 2) for in-seat food services order taking (often-called mobile concessions). Both leverage the mobility factor, but in slightly different ways.

Access Control

When a fan comes to the gate and presents a ticket, that ticket gets scanned into a hard wired turnstile connected to the wired network. An application for back end validation then confirms that the ticket represents an actual purchase, thereby minimizing counterfeiting. The access control system can also help track arrival points and volume, allowing for better dispatching of personnel and services. That same data can be used to identify when the opening of additional gates is required.

In urban areas, mobile ticket scanners can also allow for flexibility in the moving or increasing of venue gates; the data can be used to alert the venue that a street needs to be closed off in order to create an additional gate and an extended fan area which, in turn, can function like a street mall populated with food services and hard goods sales that support the event experience and generate additional revenue. (When the event is complete the street space is returned to the municipality.)

For an outdoor event, like a concert, or a less urban venue, mobile ticket scanners can provide more flexible points of entry. The application can be used to help establish rings of security screening to address crowd density. In this case, movable turnstiles may be employed or handheld mobile devices can scan tickets.

Clearly, an access control application used with a robust Wi-Fi deployment can increase productivity and provide flexibility in terms of crowd management.

Mobility for Concessions

Mobile concessions can take a number of forms. A movable kiosk rolled into place for an event, and a refreshment wagon with wireless point of sale terminals are both examples of nomadic concessions. That means they are not fixed in place during non-event days nor are they continuously mobile during an event.

An example of a truly mobile concession would be a deployment where handheld scanners are used by staff for taking orders directly from the fans. Such orders would be electronically transmitted to the food preparation area where the food would be made ready for in-seat delivery. Delivery service already appears to work well in the premium seats, clubs, and suites. This type of application combines a higher level of service for all fans with increased productivity and efficiency from the concessioner perspective.

First, only the food requested is prepared, allowing for less waste and better use of kitchen staff. Secondly, only the food requested gets carried which, again, makes for less waste and means that every delivery is associated with a revenue-producing event. Third, fans do not need to leave their seats and possibly miss out on the action on the field. In many venues this could be presented as a premium service that improves the event experience for all fans.

Another option for venues provides a web-based enhanced self-service alternative. In this approach, the fan would use a web application on their smart phone or tablet to place the order and pay, then pick up the customized order at a concession stand window without having to wait on the general service line.

Working with Hand Held Scanners

You've now been introduced to just a few of the benefits of using hand held scanners as your point of sale delivery mechanism. There are many other uses for these devices, which are offered by a number of vendors such as Motorola and Orange. They are customized by the concessioner to display the venue specific menu content and provide back end system interaction. Your Wi-Fi design has to provide them fast, robust, seamless mobility to enable this revenue and service stream.

Meeting the needs of highly mobile business applications, especially ones that directly service the fans, usually requires constant availability plus fast, predictable roaming between access points.

It can be very desirable to dedicate a unique channel to the services using these devices. Think of it as the Service Critical Wi-Fi layer. This RF layer can be reserved for all venue-owned or controlled fixed function devices, such as hand held scanners. These devices can have their network parameters tuned for the venue before they are placed into service. In this way, the venue ensures there is sufficient network capacity for vital, revenue producing services and the associated devices are set to leverage capacity.

The tuning process starts with the creation of a unique System Service Set Identifier (SSID). This unique network can then be constrained to the area where the mobile service is offered. In the case of access control ticket scanners, this would be the entrance gates while, for mobile concession scanners, it may be the premium seats or other designated concession areas. (It is very common for a concession service person to be assigned to a specific geography within the venue, like a section or a club.)

Setting the devices like this supports that service person's productivity by allowing them to take advantage of a network that can virtualize its presence as the device moves between access points supporting its function.

Creating that mobility layer then becomes a combination of deploying a unique SSID, pruning it for a bounded services area, deploying the program on a dedicated channel reserved for that application, vitalizing the RF environment and tuning the devices for your maximum advantage.

Venue owned and controlled devices can be used by your IT department or consultant to test the network and optimize the device network parameters. Doing so allows you to select the optimal security profile, usually WPA2-PSK, as well as to assign or reserve a predetermined IP address and select a frequency band, optimally 5GHz.

A well-designed mobility layer should optimally present itself to these devices as a single entity. The concept is similar to the way a cellular network presents itself to the handset. A Wi-Fi mobility layer would virtualize the individual nature of the access point for a single presence representation. Once the abstraction of a single access point is presented, it would then be the role of the network to guide the client to the best possible radio. This is often known as referring to a network in controlled seamless roaming.

While tuning an environment for mobility, the visibility of station roaming events becomes very important. In addition to per event reporting, the system should be able to store good documentation of the physical environment. It is recommended that access points be provided with operationally recognizable names to assist with this process. Using quickly identifiable AP names that reflect the level, section and grid line, as well as the operator involved, is highly recommended for easy verification of the system, provided station roaming details.

For example, an AP in the second level, blue seating area mounted in section 126 would have a prefix of 2-blue-126 and a suffix to pinpoint its map level, section and gridline coordinates (2-44-11) for identification on a floor plan. Logically, when verifying a system initiated roam, you could then observe a handoff between 2-blue-126 and 2-blue-130.

During the Game Day tuning process, getting access to crowded and active areas with measuring equipment may not be easily accomplished. For the mobility layer, it is usually beneficial to have remote instrumentation, which includes spectrum analysis and packet capture capabilities.

The Working Press

For the working press, attending the event is likely not just a job, but a passion. They know the game well, arrive early and stay late. They are the voice of the venue to the fans and some of them travel to follow the team. In other words, the working press is an important element in supporting the business. In turn, they should be perceived as and supported by the venue IT staff as if they were part of an operational department — only not privileged to internal network access (like employees or contractors).

The working press' observation point can vary from on the court to special areas in the stands or a dedicated Press Box. In their designated area there are usually seats assigned by media outlet. Electricity, phone and hardwired network drops are usually provided. Today, of these three utilities, only electricity is regularly consumed.

During an event, it is common to see writers or reporters with three enabled devices, each being used for a separate purpose. People use their phones to stay in contact with their organization, peers, family and friends. They most often use their tablets for streaming the game or a related event. And,

Field View of a Press Area

they use their laptops for looking up Internet data, writing and submitting their work. Your Wi-Fi design needs to accommodate multiple devices per user in a high-density seating area, assuming high network load expectation per device. There is a limited expectation of seamless mobility as members of the press predominantly stay within the media area.

The devices they carry vary greatly in age and capability. Phones and tablets may be personal devices while laptops may be their own or supplied by their employers. This presents an additional consideration as an employer-supplied laptop is often locked into its organization's security domain. The writer in question may have very limited privileges on such a device and administrator access to the device is rarely available to the individual. While supporting the press, assistance for device specific anomalies have to be addressed within this limited ownership context.

Likewise, full assistance to improve connectivity for older hardware, network drivers or network cards cannot be readily offered. For the increasingly small number of people in this situation, offering a hard line network drop and a plain old telephone line (pots) is a viable approach.

The good news is that, on average, newer equipment is being brought to the venue by the working press. This is evidenced by the fact that there is increasing usage of the 5 GHz Wi-Fi band within most press areas. In comparison, general fan areas experience more of a mixed usage by band. This observation indicates that your Wi-Fi design of the press area must include more 5GHz channels at wider channel widths to better service this user population.

Additionally, providing excellent service in the area will occasionally involve dealing with an anomaly in the network or in the devices. Network anomalies, such as interference, can come from the usually adjacent television broadcast or other audio-visual sources. One proven method for supporting this area is providing a dedicated spectrum sensor for monitoring channel utilization and non-Wi-Fi interference.

The technique of 'showing a user' (conducting a three points of view test using devices running at the same Internet speed) is also very effective in highlighting network versus end station performance conditions. If all three test devices perform well, and the user's device does not perform well, you can shift your focus to what is different with the user's device as opposed to what is wrong with the network.

Concierge Service for Premium Seats

Most venues have areas designated for premium seating. They are courtside, ringside, behind home plate or 50th yard line field level. Luxury suites, Sky-boxes, function rooms, private meeting rooms and sponsored clubs fall into this deployment category as well. Many times these luxury suites or boxes are available to patrons at high prices, raising expectations about service as well as view of the event. These suites can accommodate fewer than 10 fans or upwards of 30, depending on the venue.

In these areas you will find the most affluent among the dedicated fans. The luxury boxes will often host business people enjoying the event with clients and their families. These fans have high expectations of good service.

Providing a high quality experience in the premium sections should be one of the first phases of Wi-Fi deployment, usually undertaken right after Press Box deployment.

These areas usually have a higher concentration of more powerful Wi-Fi devices than the general seating area. Private space plus easy access to electricity allows for more convenient and more often use of a laptop or tablet computer. As you might expect given the affluence of attendees, these spaces often experience the presence of mostly new technology devices.

From a Wi-Fi design perspective, there is a need for modest capacity with minimal need for mobility; when taken together, these requirements allow for a straightforward design. For luxury boxes the deployment of one access point for every other suite is common.

The club areas usually do not want the access point exposed. Since they often have dropped ceilings, access points can be mounted on structures above the ceiling. When possible, it is recommended that an unobtrusive external antenna be placed below the ceiling tile.

Concierge service is usually available from the foods services and security staff. Given premium fan expectations of special service, it is recommended that a Wi-Fi manager representing the IT staff be assigned to this area. He or she can assist fans with their devices and quickly report problems in service.

In Venue Club Room Used for Meetings

In the sponsored clubs, you should probably design Wi-Fi with evenly spaced access points using a guideline of one access point for every 5,000-8,000 sq ft. This allocation can be further refined by assuming one access point per 100 fans. If a function room will be used for high capacity Wi-Fi functions, it may need to be designed similarly to the Press Box. That is using the metric of one access point for every attendee. For variable use rooms, temporary access points should be brought in as required to supplement the standing infrastructure.

Fan Access - The Seating Bowl

The basic bowl configuration of large spectator venues has evolved from ancient amphitheaters, which were open-air, round or oval structures. In fact, the concept of a high capacity entertainment venue with a free standing central arena, an agricultural-like terrace and tiers of concentric seats in a freestanding structure has been with us since long before Rome's Coliseum held some 50,000 people.

Varying Configurations

Today, different sports require playing areas of different sizes and shapes. Some stadiums and arenas are designed primarily for a single sport while others can accommodate several different types of events. From a Wi-Fi physical design perspective, a major challenge for venues is that the seating stands are necessarily set back a good distance from the playing field; additionally they are steeply pitched vertically.

When there are stands all the way around, the stadium takes on an oval shape. Some stadiums do not have seats at the ends. When one end is open, the stadium has a horseshoe shape. Three major configurations (open, oval and horseshoe) are common, especially in American college football stadiums. (Rectangular stadiums are more common in Europe.)

Stadiums have evolved over time, with many having distinct and very different seating configurations on various sides of the stadium. These sections are often all of different sizes and designs and have often been erected at different periods in the stadium's history. In many cases, earlier baseball stadiums were constructed to fit into a particular land area or city block. This resulted in asymmetrical dimensions for many baseball fields. The old Yankee Stadium, for example, was built on a triangular city block in the borough of The Bronx, a part of New York City. This resulted in a large left field dimension but a small right field one.

In the United States, baseball and football are the two most popular outdoor spectator sports, so it is not surprising that there are a number of football/baseball multi-use stadiums. The requirements for baseball and football are significantly different, which can be problematic because, at times, their seasons can overlap.

Since the 1990s, there has been a trend toward the construction of single-purpose stadiums like the medium sized ones used for ballparks. The Miami Marlins Ballpark, built in 2012 for Major League Baseball, is an excellent example of a moderate size venue. It has 37,000 seats, is outside of downtown and has amenities including a retractable roof and air conditioning.

In several cases, such as in Philadelphia, the football stadium has been constructed adjacent to a baseball park to allow for the sharing of parking lots and amenities. With the rise of Major League Soccer and the construction of soccer-specific stadiums there is growing interest in the sports complex concept.

Antennas and Algorithms

Acoustics is about geometry. RF, like sound, travels by line of sight. Stadium designers often study acoustics to increase the noise of the crowd's voices, aiming to create an energized atmosphere. The purposeful acoustics designed into outdoor stadiums makes designing free space reflection dramatically different than when working within a closed structure. RF will travel much further and clearer when it is transitioning an open area. Technically there is very little "free space path loss" in the open area of the stadium.

A Wi-Fi radio in one section can be heard by a client station on the far end of the seating bowl while the same radio cannot be heard one level above or below.

As touched on previously, just because a station can hear an access point does not imply the station can transmit with enough RF energy to be heard by the access point. The Wi-Fi conversation, therefore, becomes one about symmetry and an access point's receiving radio sensitivity. You will need to use rejecting reflections positively — IEEE 802.11n technology. Employing a previous analogy, symmetric coverage means not designing your RF environment as if you were a lighthouse whose beam can be seen for miles. Rather, make sure that the flashlight can be seen. When you turn down the transmit power, you can listen better.

Using narrow beam antennas is one way of minimizing the number of clients an AP sees so it can focus attention on a specific section of clients. This hard wired method is similar to doing manual work. It is not to be used alone but as part of a broader approach. Think of it this way —productivity is amplified in human effort by better use of the available resources, such as time. The same holds true for stadium class Wi-Fi design; algorithms will be more productive than antennas. The situation is also analogous to happenings in the Industrial Age, when the steam shovel proved more productive than manual labor. It is merely a matter of using your back or your brain.

Sectional Use of Access Points and Antennas

Larger outdoor venues, like stadiums, have an advantage over indoor arenas because they have the ability to use the distance between sections to reuse the spectrum. Spatial reuse is the term applicable for simultaneous transmission on the same channel on a per section basis.

Systems can be designed for special reuse if you know how to identify the interference region of the neighboring access points. This is similar to the way many people are able to talk simultaneously in a crowded restaurant or coffee house. They may lower their voices when close to each other and just listen to their intended conversational partner. They are able to ignore the specifics of the other conversations, which get treated as undistinguishable background noise. Instead of hindering an individual conversation, that background noise ensures a basic level of privacy for guests who are talking to each other. Many restaurants replace the 'din' with music for the same purpose.

Most access point mounting will be on overhead concrete overhangs to accommodate the steep nature of the terraced seats. Three AP mounting antenna directions – forward, down and backward can be accommodated from one mount point.

Overhead mounted access points have additional serviceability implications. Each AP in an area can service up to 200 fans, creating the need for a high-speed connection to the wired network. Since the access point will be exposed to outdoor environmental factors, it will usually require the installation of a ruggedized access point or the use of a NEMA approved access point enclosure. Of the two choices, the enclosure is usually the most flexible and cost effective alternative.

Additionally, the required high quality Ethernet cable will most likely have to be installed inside a metal conduit to meet environmental and safety specifications. A scissor lift or boom will have to be used to access the device. And, the need for long cable runs, inside conduit penetrating or working around a concrete structure make the cost of each access point drop costly. Think of this as the Concrete and Conduit Problem.

Working Within The Bleachers

The other high-density spectator-seating configuration may be referred to as the 'bleachers.' A key feature of bleachers is that they are typically uncovered and unprotected from the sun; hence the name, as the wooden seats were 'bleached' by the sun. Bleacher structures vary depending on the location, but most outdoor modern bleachers have aluminum benches over a steel support structure.

Bleachers range in size from small, modular aluminum stands that can be moved around soccer or hockey fields to large permanent structures that flank each side of a football field. In arenas and gyms, bleachers can be built-in so that they slide on a track or on wheels and fold in an accordion-like stacking manner. It is not uncommon to see football bleachers that rise hundreds of feet into the air. Football bleachers are commonly made from concrete or aluminum with concrete footings or superstructure underneath.

Though many stadiums offer only bleacher seating, in those that offer both seats and bleachers, the bleachers are typically in less desirable locations and/or have lower ticket prices, giving the term "bleachers" a connotation of lower-class seating. There can be a sense of pride that comes from sitting in the bleachers; in New York's Yankee Stadium the fans sitting there are referred to as "the bleacher creatures." A bleacher is known as a grandstand when it contains VIP seats.

Bleachers are different from other seating areas when it comes to Wi-Fi deployment as access points can be mounted under the seats while above seat mounting can be employed for the antennas. (Traditional seating would require concrete penetrations and re-pouring to accomplish this approach.) This is desirable as close pairing of a fan's device to an access point makes for symmetric coverage. The risk of damage to the partially exposed antenna wires is offset by the inexpensive nature of antenna replacement. You may want to consider a handrail mounting of antennas to avoid blocking the fans' view.

Hey Access point "Can you hear me now"

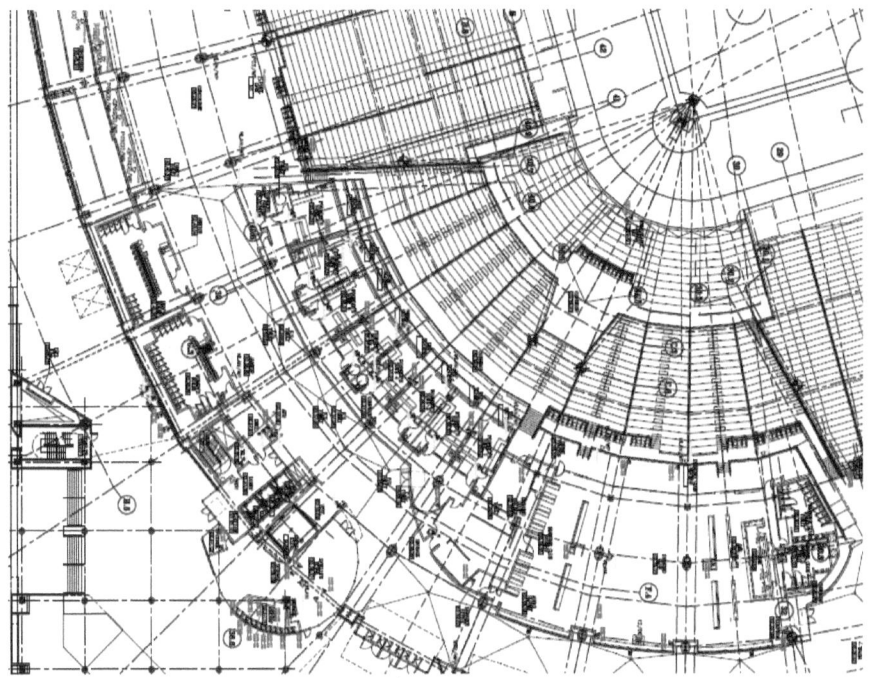

Alternate Events Flexibility

As mentioned above, many venues have a primary function while a few have a fixed function. Arenas often serve basketball and hockey, with the hardwood and ice exchanged as needed. Additionally, as a treat to many fans, hockey is occasionally played in open-air baseball stadiums.

Since the 1960s, when The Beatles played Shea Stadium in New York, stadiums have also been used as live music venues. From the 1980s on, rock, pop and country music stars including Madonna, Bruce Springsteen, Taylor Swift and Lady Gaga have undertaken large-scale stadium based tours.

Over time, the increased power of amplification and sound systems allowed for the use of larger and larger venues. Today, music concerts are held both indoors and outdoors based on venue availability, the expected size of the crowd and other needs of fans and promoters. Smoke, fireworks and sophisticated lighting shows are often staples of arena performances.

Flexibility is key to a successful Wi-Fi deployment in a stadium.

If the fans are in the seats or on the field they expect to share the moment with family and friends. They expect the Wi-Fi network to keep them connected. This typically calls for the extension of the network into places where a direct network connection is not usually available. Floor boxes, tripod mounts and stage light trusses all become viable mounting points for event network data drops.

Occasionally, there is a request for network coverage in a no-data facility. A common component for this requirement is an access point to access point RF connection. The primary function is to provide a backhaul of the remote access point onto the wired network. One access point serves as the gateway to the wired network while the other is the fan servicing endpoint. The setting up of a number of such links is commonly referred to as a mesh network or a mesh hop.

Wi-Fi is by nature a half duplex protocol, whether you are sending or receiving data (but not both). Each mesh hop will, therefore, have only half the bandwidth of the previous hop. For example, if the data rate between two mesh end points is 300megs, the maximum available throughput is 150megs. Extending this link to the end station drops the maximum to 75megs. In many cases, only one mesh hop is needed to accomplish on the field fan coverage.

When designing a sports or entrainment Wi-Fi network, upfront consideration should be extended to the need for temporary expansion. In addition to covering the variation of known configurations, event APs with backhaul capability allow for on demand coverage outside hospitality tents (sometimes called the Fan Zone), parking lot festivals or other events that extend 'beyond the walls'.

Flexibility considerations can be extended to remote coverage of affiliate teams, temporary branch offices or secure in-house support staff access.

Chapter 7

The Control Plane

The International Standards Organization has provided an excellent layered frame of reference for looking at network systems. This model characterizes and standardizes the internal functions of a communication system by partitioning it into abstraction layers.

This method of categorization can be extended to the working components of Wi-Fi as a system. In the model, a functional layer serves the layer above it and is served by the layer below it. For example, a layer that provides error-free communications across a network provides the path needed from applications above it. At the same time, it calls the next lower layer below it to send and receive packets that make up the contents of that path. In the case of Wi-Fi, the physical layer is the RF frequencies. These are controlled by the radios in the access points.

The processor in the access point controls the Wi-Fi media access control protocol 802.11. This allows the traditional use of the OSI seven layer model called the Data plane. There is also a plane of control that coordinates the behavior of all access points as the function as a system. To the side of these layers is a management plan that touches all layers, managing configuration and collecting data about the networks usage and capacity.

In some designs, the control function only implies control point administration of the static AP settings. Stand-alone access points need to be configured separately by an administrator. A centralized configuration model relieves you from having to configure each access point individually. The idea of centralizing the administrative functions using a controller took wireless out of the sphere of the home and into the corporate administrative domain.

An element minimized in most designs is the active real-time control plan relationship between the access point and the controller. Together the two manage end station data flows. This relationship is important. Active Control of well periodic network function, such as channel and power settings, are a step in the right direction, but it falls short of what is required to maintain a stadium scale environment. The key element is the amount of client side control that can be exerted. A design consideration is that software cannot be deployed on the client side to facilitate active client engagement. The only methods available are the Wi-Fi standards already built-in to the devices. It is in the control plane where the network can create a global topology, adjust handoff characteristics and minimize the effect of a slow data rate or poorly associated, sticky clients.

Service Control

Wi-Fi is becoming a requirement of in-venue video content distribution. Two multicast variations are emerging: downstream and upstream. Downstream distribution of instant replays and multiple camera angles will use a multicast design to allow fans to select a "channel" on which to watch using their small screen devices. Business users will, over time, have the need to use upstream multicast traffic to share content with departmental video displays and printers. The mechanism for deploying multicast video control services will use control points to minimized and isolate traffic.

Zone Coverage

In a stadium environment multiple control points will be established to sectionalize the control function, to create fault resilience and to provide expanded capacity and differentiated services. The function can be distributed to multiple points within the venue as defined by geographic section (such as the upper bowl, the blue section) or function (such as premium seats).

In low capacity environments, this control plane can be effectively distributed among the access points, or it can reside in the Internet cloud. In stadium deployments, the control function is optimally gathered in fixed function devices usually called controllers.

The key element in this design has to do with using a network structure such as virtual land (VLANS) or IP routing to facilitate mapping of access points to their control points. This allows multiple controllers to be deployed into centralized computer rooms or physically distributed in wiring closets.

The control point can be in line to the flow of data traffic, creating a centralized network entry point, or it can be just a coordinator providing the access points a central authentication point, central client topology, client control and inter AP air coordination functions.

It is critical to understand the need and function of the control plane in helping to refine and optimize the services the network can offer. It is, therefore, a key design element if you are to have the flexibility to deploy and change the arrangement over time in response to business requirements.

How would you control the air for 5000 active bloggers using the network by in one function large room? Where would place "the brain"?

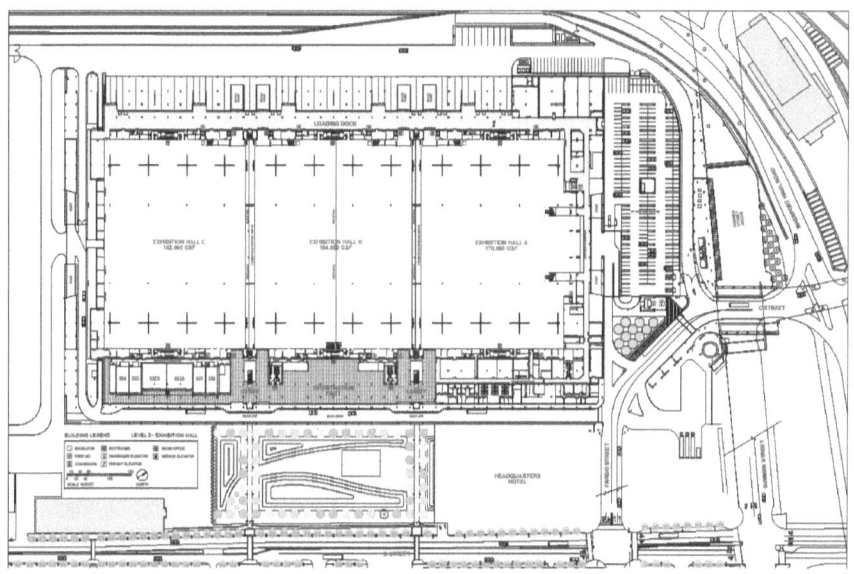

Hynes Convention Center, Boston

Chapter 8

Game Day Monitoring

A game happens only once then it becomes history, so there is only one chance to deliver a high quality Wi-Fi network experience. This makes monitoring extremely crucial on many levels. It is critical to include a monitoring and reporting layer into your design. First, look to verify the system is performing as designed. Second look for feedback to refine the design. Third, look for changing user utilization patterns to provide input into refining the design. While handling these tasks, you cannot ignore responding to problem reports. One of the practical steps in designing the network is to build in a problem response plan.

Assembling the Firemen

There is an analogy to be made between fire department and ambulance corps stationed at sporting events to monitor the health of the athletes, and the Wi-Fi network support staff. They are not always needed. That can look like an inefficient use of human capital when nothing goes wrong. However, their presence looks like brilliant planning when something does go wrong and is quickly and professionally addressed. Not every event requires the same level of safety staffing and response planning.

There is also flexibility and staff hierarchy. The same is needed in your IT staff. Starting with the lowest level position, volunteer Wi-Fi coaches could be recruited from among your fan base to assist other fans with their Wi-Fi usage. These coaches could help adjust the settings on devices, add a password or select a network name. Consider buying these volunteers special identifying tee shirts or blazers, as they need to be as noticeable as Ushers.

The next level of IT support could be comprised of mobile network technicians who could interact with facility staff, adjusting or replacing devices, or request a diagnostic procedure be run on a particular area of the venue. In our analogy, these technicians are the bulk of your first responders. They are your alert system.

Moving up in the hierarchy, you could also have more senior technicians authorized to make configuration changes to network infrastructure devices such as routers, firewalls, switches and wireless control plane devices. These people represent your paid firemen, the ones who populate the firehouse.

For major events a command response center can be staffed with vendor consultants and project managers. In our analogy, the paid firemen would be replaced by network technicians from the IT group and on-site vendor support people. The command center would have remote monitoring backup and escalation.

You will want to create a plan in which each level in the hierarchy is empowered to make increasingly more serious levels of changes. One important caveat: you will need to prevent possible over reaction to a situation. This can be done by not allowing first line information to be delivered to top-level support staff without verification.

Another key similarity between the fireman analogy and your Wi-Fi support system is the need to have a planned alert, escalation and response communication strategy. A prominent Stadium IT director told me "having the Wi-Fi fail during a game event would be a Disaster." While not a disaster in terms of a hurricane or earthquake it is an emergency, a Wi-Fi outage would definitely shake up a venue. Considering the user expectation for always available Wi-Fi, and the benefits of having a well written and well practiced time based Emergency response and recovery plan, a well crafted plan should be mandatory.

At any one time, there could be multiple incidents happening. Creating a network command center can help prioritize them. This command center can also dispatch the appropriate resource to the proper problem.

Here's an example of how this recommended IT structure works. Near the end of a high profile baseball doubleheader a report from a marketing executive is sent to the stadium IT director to say that the Wi-Fi is 'out' in the premium seating area, and the concessioners are taking in-seat food orders on paper tickets instead of using the Wi-Fi ordering system.

Without a structured response process, all network technicians and all available vendor support staff might have to be rushed to the problem site. Instead, using a structured response plan, the vendor technician can investigate the management station and control plane consoles. The network manager contacts the concession manager to identify the specific area and devices involved. Then a network technician is dispatched with diagnostic equipment to the specific section. Within moments, the problem is identified as isolated to a single hand held device. The device is inspected and found to have a dead battery due to prolonged use during the double game. The battery is replaced solving the problem. A report is quickly made to the marketing director by the IT director about the soundness of the network and the resolution of the problem.

Network Management — One 'Pain' of Glass

Having an overall view of the system is helpful in monitoring the environment for capacity planning. It is also helpful in correlating data for forensics or failure root-cause analysis. To consolidate a view of the network use standard reporting interfaces through an enterprise management platform.

Most venues have a variety of best in class products from many manufacturers in their network, including routers, firewalls, switches, servers, disk arrays and wireless components. Interoperability standards have emerged to consolidate the proprietary vendor specific network management protocols.

One key to successful management of innovative products is the employment of standard management protocols such as the Simple Network Management Protocol (SNMP). Together with standardized System Event logs and Syslog, SNMP allows consolidated management programs to collect data from best of breed solutions. Event monitoring then has two components' polling systems that create overview observation and alerting functions which can be provided by consolidated environment view programs and real-time device control systems.

The monitoring staff starts with the understanding that Wi-Fi operates in an unlicensed RF frequency band. No one has a 'legal' write to control this spectrum. For game events in private venues, you can exert private domain control. So, ensuring the system operates as designed implies optimizing control plane elements of the Wi-Fi solution while eliminating items that interfere with system performance. Think of this two-pronged approach as tuning 'in' the good and tuning 'out' the bad.

Tuning In the Good

Event monitoring of the Wi-Fi control points then has two main components: polling systems that create overview observation and alerting functions.

In the previous section there is further discussion about a consolidated environment view and capacity planning. Consolidation platforms, while very important, are quickly bypassed during live event monitoring. When troubleshooting, engineers will quickly move to the vendor specific graphical element managers, then the device specific command line interfaces.

You can separate the real-time radio functions and client station coordination functions. The aggregation of all the station data from all the controller points into a central place for capacity planning is the function of the management plan

Monitoring then becomes a management operation while tuning is considered adjustment to the control plane.

Anticipate and Engage

Compare your design work to reality and make adjustments. During the tuning process you must be aware of what the system was designed to accomplish and know which redundancy and backup systems are in place. This is of utmost importance in Game Day monitoring.

Critical business systems need specific metrics, and need to be monitored first. How many hand held scanners will be deployed? Are they all attached to the network? Are they connecting to the set of access points? How many people are expected in the Press Box? Are the luxury suites all receiving associations at high data rates?

In certain designated areas, there will be a well-known set of devices operating for business critical functions. It is important to place packet capture sensors and dedicated spectrum analyzers in place. During the event, it is best practices to provide stakeholders with access to internal IT so they can report problems.

If reports come to the help desk during the event, before even dispatching a Wi-Fi technician, you can turn on these sensors to record detail events for Tier 2 analysis and vendor support if required.

Monitoring the stadium bowl for fan access requires a broad view of system operations. The Game Day monitoring plan defines visualization components such as placing APs on maps, defining geographically specific AP names, having control plane groups, AP groups and verification of access point locations.

Among the key design elements to be monitored are:

- Station per radio

- Network bandwidth utilizations

- Distribution of users among available channel layers

- Packet loss

- Statistical station distributions

- Usage per band

- User migration patterns

- RF retries

- Top talkers

- Network hot spots where there is excessive loading of wireless clients onto a specific AP

Tuning Out the Bad — Interference and Rogues

In addition to the performance of system components under load, external factors can contribute to the fans' perception of the systems' performance. These external factors can be separated into Wi-Fi and non-Wi-Fi interference.

Each game will bring in a new group of fans and a new group of radio and TV broadcasters. The second part Game Day monitoring and tuning is removing Wi-Fi network interference from these sources.

Start with a Policy

Fans, press, broadcasters and stadium staff do not usually try to sabotage event Wi-Fi deliberately, but some may do so unintentionally. Interference can be produced by many sources that utilize wireless service to control their equipment (such as wireless video cameras or a blimp dropping coupons).

Your biggest concern will probably be non-Wi-Fi-compliant continuous broadcast devices such as wireless cameras. Older analog wireless security cameras are mostly fixed frequency devices, with higher potential for affecting Wi-Fi APs in overlapping frequencies. These analog cameras can have a 100% duty cycle, which is even higher than that of a microwave oven (50% duty cycle).

Make it clear to all visitors that the venue has the authority to regulate any and all devices that can interfere with facility Wi-Fi operations.

Set a policy for coordinating with non-Wi-Fi interference sources and you will almost be able to eliminate them.

Non-Wi-Fi Interference - Frequency Coordination

For the licensed RF spectrum assigned to broadcast radio and TV, outsourced frequency coordination can be extremely effective. It is reported that for NFL games you should coordinate from 400-500 frequencies during normal games and over 2,000 frequencies and over 10,000 RF devices for the Superbowl.

Many other types of devices emitting in the unlicensed band dwarf the power and place constraints on Wi-Fi. These devices include microwave ovens, cordless phones, Bluetooth devices, wireless video cameras, outdoor microwave links, wireless game controllers, Zigbee devices, fluorescent lights, WiMAX, and so on. Even bad electrical connections can cause broad RF spectrum emissions. These non-802.11 types of interference typically do not work cooperatively with 802.11 devices.

The 802.11 protocol is designed to be somewhat resilient to interference. When an 802.11 device senses there is a burst of interference occurring before it has started its own transmission, it will hold off transmission until the interference burst is finished.

If an interference burst starts in the middle of an ongoing 802.11 transmission (and results in a packet or packets not being received properly), the lack of an acknowledgement packet will cause the transmitter to resend the packet.

In the end, packets generally get through or are lost. The result of all these hold-offs and retransmissions, however, is that the throughput and capacity of your wireless network is significantly impacted.

For example, microwave ovens emit interference on a 50% duty cycle (as they cycle on and off with the 60-Hz AC power). This means that a microwave oven operating at the same frequency as one of your 802.11 access points can reduce the effective throughput and capacity of your access by 50%. So, if your access point was designed to achieve 24 Mbps, it may now be reduced to 12 Mbps if it is in the vicinity of an operating microwave.

Some venues are in urban areas where TV broadcast trucks can be parked close the facility. When they light up their power microwave transmitters on the same band as the venue's wireless, the broadcast trucks can wipe your access point 'off the face of the earth.' This loss of throughput may not be immediately obvious to you, especially if your only application on the WLAN is convenience data networking (web surfing, for example). But, if you are supporting wireless ticketing or mobile concessions sales, interference will become a critical issue.

For the licensed RF spectrum assigned to broadcast radio and TV an outsourced frequency coordination team can be extremely effective. Unlicensed Wi-Fi bands cannot always be afforded similar protection. In most venues, the role of frequency monitoring and coordinating for the Wi-Fi network falls to the IT staff. Spectrum analysis tools need to be part of your planning and monitoring tool kit for any venue.

Spectral Analysis Tools Make the Invisible Visible

Similar to a medical X-ray there is a vast amount of information in the pictures, a similarity that may take a networking equivalent of a radiologist to interpret.

Even to the untrained eye -- Nothing Good is Happening Here

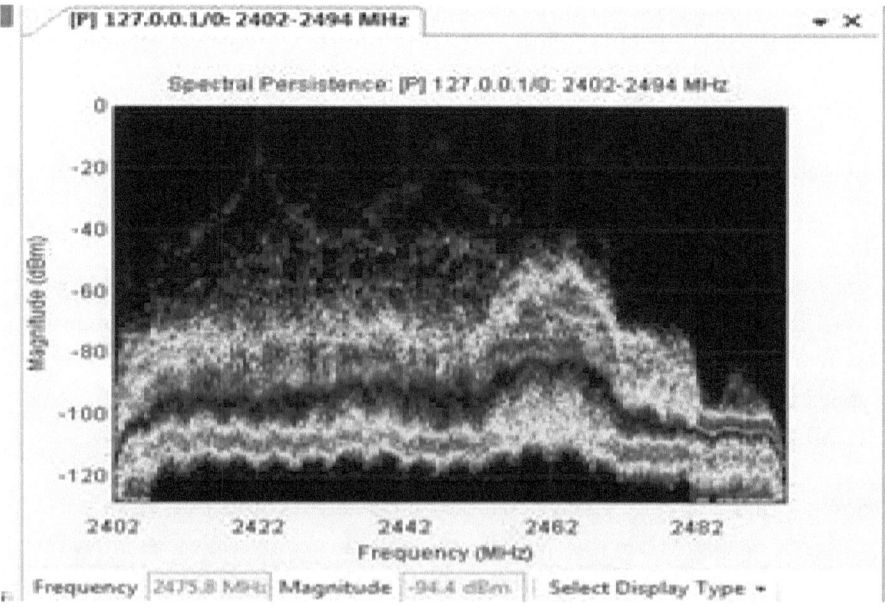

Interference on all channels – via BandSpeed spectrum manager

Interference on One Channel

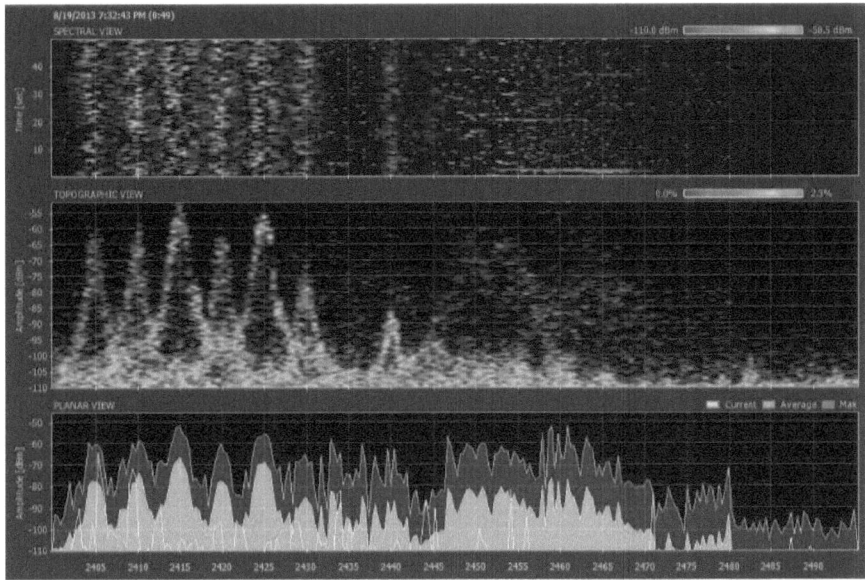

Channel 1 is being dominated by an "Interfering" non Wi-Fi device

Graph via band speed –BandSpeed spectrum manager-

Wi-Fi Interference

A 'rogue' Wi-Fi access point is any access point that is not under the control of system administrators. The key to managing Wi-Fi interference is to prevent any rogue access points or rogue equipment from attempting to operate in the same frequency as the stadium Wi-Fi network.

The best defense against such rogue wireless networking is to prevent the wrong devices from coming into the stadium. Unfortunately, you can't stop every situation from happening. For example, you can't stop a member of the working press from bringing in a laptop, as the device is essential to accomplishing their job. Yet, laptops can be problematic if their owners try to create their own private Wi-Fi networks. Anyone who enters the facility with a laptop has the ability to become a rogue by going to ad hoc wireless networking mode.

Use spectral analysis equipment for detecting potential problems, starting before the game and continuing throughout. Have a plan in place to identify an interfering signal and employees empowered with the authority to remediate the problem.

Tracking down uncoordinated and rogue devices is essential to helping maintain good signal and data flow. Once located and identified, the individuals or clients who are out of compliance must be asked to shut down or move to an acceptable Wi-Fi channel.

Channel Utilization

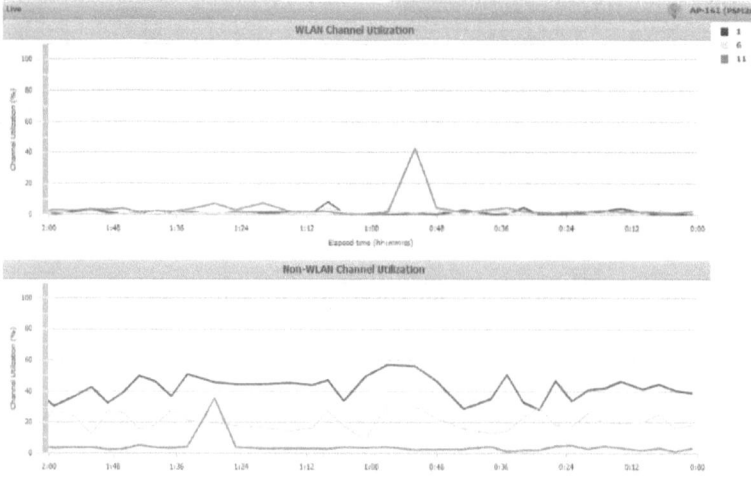

Graph via BandSpeed spectrum manager

Frequency coordination is the most effective way to guarantee the proper operation of all devices at a particular event. Taking the time to plan how to use the available spectrum will have positive results for everyone. If everyone coordinates at a particular event, then everyone can be protected by the frequency coordinator.

Chapter 9

DAS for the Carrier, Wi-Fi for the Venue

"If you do not change direction, you may end up where you are heading."
— Lao Tzu

The basic principle of RF distribution is similar to the one applied to a Distributed Antenna System (DAS), which is primarily used for cellular telephone service, and Wi-Fi in your venue. Just as a washing machine and a dishwasher both use water, electricity and plumbing, a DAS and Wi-Fi share similar characteristics with significant differences in operation and purpose.

The two systems are not interchangeable. To continue the analogy, a skilled electrician and plumber working together can install both types of appliances, but that doesn't make the appliances alike. They serve different roles and have features specific to their application. In fact, in larger environments, such as hotels and hospitals, the two appliances serve different cost centers and are primarily operated and maintained by different staff. Their use and management requirements do not align.

Similarly, at a higher level, both appliances can be managed by the same reporting mechanism: the water and electric bill. However, this single report has a limited use in tuning or managing either platform. You would not necessarily select the same vendor to operate your laundry and kitchen.

In a venue, the primary function of a DAS is to provide better cellular network service. It is operated by a network offering cell plans, such as Verizon, AT&T or T-Mobile. The DAS installation derives its value from the billable plan minutes it helps generate. It is more cost effective than bringing in Cell Towers on Wheels (COWS) and reaches into more places than COWS does.

Primarily, the Wi-Fi network is used to carry high-speed data. When designed well, it has many of the good characteristics of a cellular network. It should keep control of where the client is assigned. It should load-balance clients among antennas, provide a level of fair access of resources between the clients and provide detailed client network state records.

However, the two technologies are implemented differently. One of the key underlying differences is that Wi-Fi devices are more diverse since they are available from a wider variety of sources. The only common thread between Wi-Fi devices is the IEEE 802.11 standards they support. Cellular network devices, on the other hand, are primarily offered through the carrier providing the network. Cellular devices have carrier specific software which if fully tested for specific network compatibility. The phones are locked to a network; a Verizon phone does not readily connect to the T-Mobile network, for example, without a significant penalty.

The organization operating each of the two environments usually has drastically different needs. The DAS can assist carriers when competing with each other for customers. It does not necessarily drive value for the venue. Wi-Fi assists the venue operator by creating a fan experience through increased services such as in-seat food service, flexible ticketing and gate access, network access in corporate function rooms, premium service in suites, network access in sponsored clubs and increased internal staff productivity.

Given these differences, the DAS and Wi-Fi network need to be owned and operated by different organizations. It is not in the carrier's interest to drive the venue experience, just as it is not the venue's IT staff's responsibility to drive billable cell phone minutes.

In the race to provide enhanced content delivery for mobile devices, there will be improvements to the cellular network that will be reflected in the carriers' DAS offering. Just as the analog dial up line progressed in speed from 9.6 to 56k to 115kbps and beyond (while LAN technology moved from 10mbs to 100 mbs to 100 mbs) both carrier data networks and Wi-Fi will evolve. For the foreseeable future there will remain an importance difference in speed and capability between carrier data rates and Wi-Fi.

When speaking about mixed DAS and Wi-Fi content I was reminded by a seasoned stadium IT professional that it is the content that matters. He told me "Jim, you would not consider placing your fine china in the washing

machine as an analogy of placing your Wi-Fi content on the DAS." He continued, "but placing your voice traffic in the Wi-Fi is like placing some clothes in the dishwasher. While you may be tempted to do it to save money, you have to very careful with the arrangement of the content when you consider mixing socks with saucers." I think he was speaking about adjusting quality of service (QOS) parameters when he said, "When you place your dishes in with your draws you will want the dishes on top – you don't want any of that draw residue dripping all over you dishes." I laughed and got his point.

In the BANG the 802.11AC chapter, we'll explore how technology delivers data gigabits over the air for video content.

Chapter 10

802.11 BANG then AC

IEEE 802.11 standards current and futures

"I love standards because there are so many good ones to choose from"

-IT professional

There is a new wave of Wi-Fi innovation expected to occur over the next two years (802.11ac), and it will move from announcement to global adoption as quickly and broadly as wireless networking did. The birth of this new technology comes out of the overwhelming need to drive productivity and improve the viewing experience of video content.

The sports conversation is all about content and content is all about video, so the driver for your adoption of this technology will be your fans. As they buy newer and more technologically advanced devices, you will require the latest wireless technology in order to meet their usage expectations.

BANG

Over the last ten years, the Institute of Electrical and Electronics Engineers (IEEE) has set standards for Wi-Fi product development and interoperability that have been adopted by all commercial Wi-Fi product suppliers. The standards were individually named by working committees and designated by suffix letters; they all fall under the umbrella name 802.11. This created a wonderful assortment of radio designations, which you may have seen as 802.11b, bg,ag,abg,gn,an,abgn. In more human terms, let us refer to all the legacy 802.11 standards collectively as BANG. So now we have BANG soon we will have IEEE 802.11AC.

This standardization has the benefit of facilitating backwards compatibility between new and legacy devices. However, there is a disadvantage in that the older devices (along with their design flaws) will continue to be brought into your venues. Having good tools in your implementation toolbox is key for ensuring that the standards you provide an enhanced experience for your business and the fans. This will mean allowing sole standards and denying others. For example, not supporting 802.11b.

Let's take a quick look at the expected time line for the launch of this new standard and gauge when you may need to deploy this technology.

AC Timeline

The IEEE 802.11 set of standards that evolved into the Wi-Fi standards began in 1997 just before the 100 Mb/s standards for Ethernet were ratified. They have been updated many times, most notably in 2005 and 2012, when there was a six-fold increase in performance delivery from the 802.11n products. Through generations of Wi-Fi progress, the fundamental foundation of the standard, the underlying technology has not changed. The standards build on each other. In 2013, we saw the first production deployment of 802.11ac networks.

Wi-Fi Generation Time Line   - Meru Networks

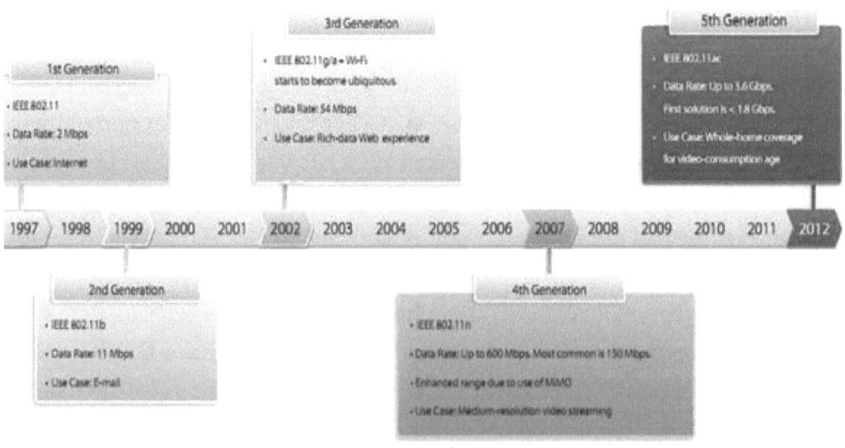

802.11ac stands on the shoulders of 802.1n

To take this leap forward, the standards committee realized more RF spectrum would be required. For several reasons, it was decided that 802.11AC would only be defined for the 5Ghz frequency band. First, the 2.4GHz band is fair saturated with non-Wi-Fi interference. Secondly, the 5GHz spectrum is currently less likely to suffer from interference, and it has more non-overlapping channels in the US than the 2.4GHz band, which only has three such channels. This offers significantly more uncongested capacity.

Thirdly, the 5GHz spectrum has been underutilized. Plus, in a stadium environment, 5GHz bands have less range than those at 2.4GHz. Rather than being restrictive, this benefits the network by lowering interference.

Wireless speed is the product of three factors: channel bandwidth, constellation density, and number of spatial streams. The 802.11ac products will be accelerated versus the 802.11n ones in each of these areas:

1. **Wider channel bonding**: Will increase from the maximum channel width of 40 MHz in 802.11n, to 80, or even 160, MHz (117% or 333% speed-ups, respectively).

2. **Denser modulation**: 256 quadrature amplitude modulation (QAM) will be upped from 802.11n's 64QAM (for a 33% speed burst at shorter, yet still usable, ranges).

3. **More multiple inputs, multiple outputs (MIMO)**: 802.11n stopped at four spatial streams, but 802.11ac goes all the way to eight (for another 100% speed-increase).

The physical layer speed of 802.11ac is calculated thusly: an 80 MHz transmission sent at 256QAM with three spatial streams and a short guard interval delivers $234 \times 3 \times 5/6 \times 8$ bits$/3.6$ μs $= 1{,}300$ Mbps.

Let's quickly look at each vector, channel width, encoding and spatial streams.

Three Dimensions of Expansion (Cisco)

Wider Channels

802.11ac adopts a keep-it-simple approach to the channelization model. Adjacent 20 MHz sub-channels are grouped into pairs to make 40 MHz channels, adjacent 40 MHz sub-channels are grouped into pairs to make 80 MHz channels and adjacent 80 MHz sub-channels are grouped into pairs to make the optional 160 MHz channels.

A Basic Service Set that is, AP plus clients uses the different bandwidths for different purposes, but the usage is primarily governed by the capabilities of clients.

	Including DFS'		Excluding DFS	
Channel size	US	EUROPE	US	EUROPE
40 MHz	8	9	4	2
80 MHz	4	5	2	1
160 MHz	1	2	—	—

Only One 160 MHz Channel is Available - Meru Networks

With the wider channel, more clients will get to transfer their data more quickly, and they can complete their transmissions that much sooner. Overall, less battery energy will be consumed, and other clients won't have as long to wait (for better QoS).

If most clients at a deployment are still 802.11n clients with 40 MHz maximum, 802.11ac will work especially well. It is allowable for two 80 MHz 802.11ac APs to select the same 80 MHz channel bandwidth; one AP puts its primary 20 MHz channel within the lower 40 MHz and the other AP puts its primary 20 MHz channel within the upper 40 MHz. This means that 802.11n clients associated with the first AP can transmit 20 or 40 MHz as usual at the same time that 802.11n clients associated with the second AP can transmit 20 or 40 MHz in parallel. New to 802.11ac is the ability for any 802.11ac client that sees that the whole 80 MHz as available to invoke a very high-speed mode and to transmit across the whole 80 MHz.

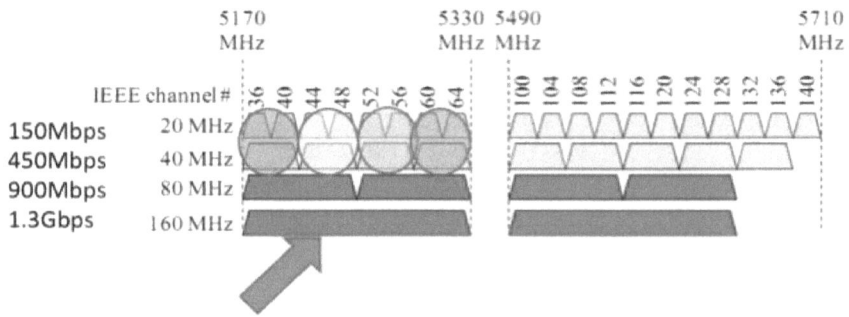

A 160MHz AC Channel = 1.3Gbs - Meru Networks

Increasing the channel bandwidth to 80 MHz yields a 2.16 times speed-up and 160 MHz offer a further doubling. However, nothing is free: increased bandwidth consumes more spectrum, and each time you split the same transmit power over twice as many subcarriers the range for that doubled speed is slightly reduced. It is still an overall win as it allows you to design for less interference plus greater transmission symmetry between the client and the access point.

Spatial Streams

Spatial "multiplexing" is a technique whereby multiple antennas separately send different flows of individually encoded signals (called spatial streams) over the air in parallel, thus reusing the wireless medium or multiplexing the signals to put more data through a given channel. At the receiving end, each antenna sees a different mix of the signal streams. In order to decode them accurately, the receiving device needs to separate the signals back out or "de-multiplex" them.

In an 802.11n MIMO environment, multiple antennae are used, both at a receiving station and at a transmitting station. In order to transmit and receive various signals simultaneously, a common multiplexing technique known as spatial multiplexing is used. Basically, when wireless signals are being transmitted simultaneously from different antennae, each signal is transmitted via a spatial stream within the given spectral channel to avoid collisions.

It is helpful here to review a single stream 802.11 transmitter.

802.11 Classic Transmitter

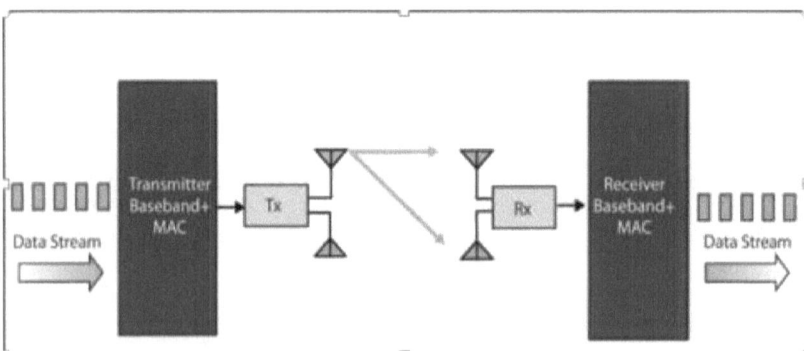

Source Wikipedia

In this scenario, only one data stream is sent from the transmitter to the receiver (represented by the orange line).

With spatial multiplexing, multiple data streams are transmitted at the same time. They are transmitted on the same channel, but by different antenna. These are referred to as antenna chains. They are recombined at the receiver using MIMO signal processing. This is represented in the diagram below with two spatial streams – an orange colored one and a navy blue colored one.

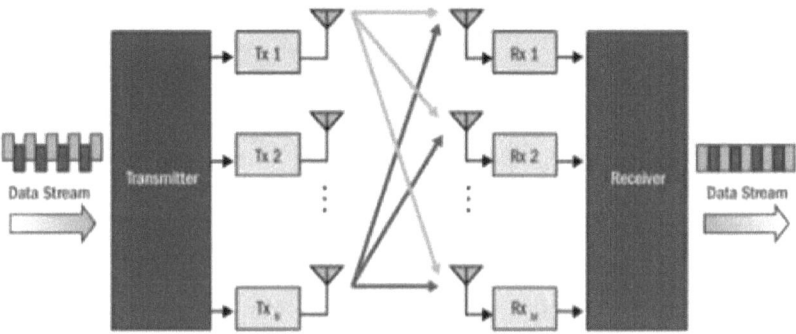

Three Antenna Chains Two Spatial Streams

Source Wikipedia

Spatial Multiplexing - Streams

Spatial multiplexing doubles, triples or quadruples the data rate depending on the number of transmit antennas. With 802.11 naming conventions, the first number is for the transmit antenna, the second is for the receive antenna and the third is the number of spatial streams. For example, a 3×3×2 system has two spatial streams.

802.11ac can have five to eight spatial streams. With three streams, 1.3 Gbps of bandwidth can be seen over the air.

Rings of Service

Bandwidth (MHz)	Number of Spatial Streams	Constellation Size and Rate	Guard Interval	PHY Data Rate (Mbps)	Throughput (Mbps)*
40	4	64QAMr5/6	Short	600	420
80	1	64QAMr5/6	Long	293	210
80	1	256QAMr5/6	Short	433	300
80	2	256QAMr5/6	Short	867	610
80	3	256QAMr5/6	Short	1300	910
80	8	256QAMr5/6	Short	3470	2400
160	1	256QAMr5/6	Short	867	610
160	2	256QAMr5/6	Short	1730	1200
160	3	256QAMr5/6	Short	2600	1800
160	4	256QAMr5/6	Short	3470	2400
160	8	256QAMr5/6	Short	6930	4900

802.11 AC Speeds – for Rings of Service   Source Wikipedia

Think of a three spatial stream AP as having three 'rings' of service based on distance. The inner ring (best SNR) may yield three spatial stream performance. The middle ring may yield two spatial stream performance. The outside ring can only support a single spatial stream. This is important when balancing service with cost as well as creating premium services is premium areas.

QAM - quadrature amplitude modulation

Constellation Density

QAM is a technique that conveys two digital bit streams by changing the modulation of the amplitudes of two carrier waves, using the amplitude-shift keying (ASK) digital modulation scheme. The two carrier waves are out of phase with each other by 90° and are thus called quadrature carriers or

quadrature components — hence the name of the scheme.

The modulated waves are summed, and the resulting waveform is a combination of both phase-shift keying (PSK) and amplitude-shift keying (ASK). A finite number of at least two phases and at least two amplitudes are used. PSK modulators are often designed using the QAM principle, but are not considered as QAM since the amplitude of the modulated carrier signal is constant.

QAM is used extensively as a modulation scheme for digital telecommunication systems. Arbitrarily high spectral efficiencies can be achieved with QAM by setting a suitable constellation size, limited only by the noise level and linearity of the communications channel.

The constellation diagrams below show the different positions for the states within different forms of QAM as the order of the modulation increases, so does the number of points on the QAM constellation diagram.

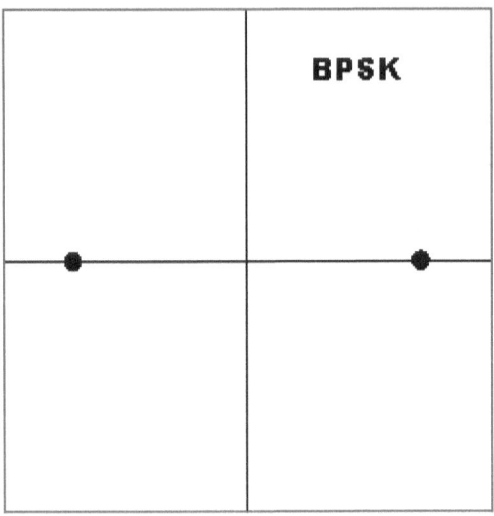

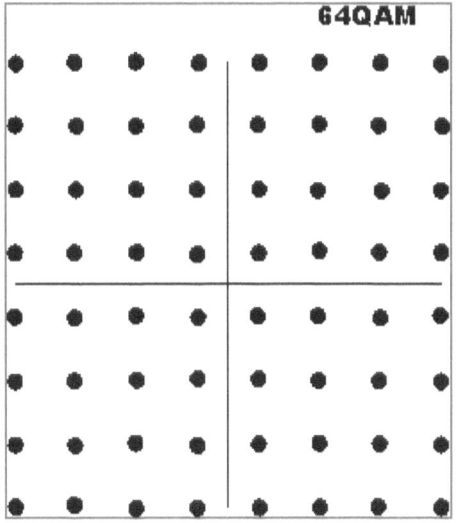

Gathering the light from many stars to shine brightly

QAM Bits per Symbol

The advantage of using QAM is that it is able to carry more bits of information per symbol. By selecting a higher order format of QAM, the data rate of a link can be increased.

The table below gives a summary of the bit rates of different forms of QAM and PSK.

Modulation	Bits per symbol	Symbol rate
BPSK	1	1 x bit rate
QPSK	2	1/2 bit rate
8PSK	3	1/3 bit rate
16QAM	4	1/4 bit rate
32QAM	5	1/5 bit rate
64QAM	**6**	**1/6 bit rate**

			Theoretical throughput for single Spatial Stream (in Mb/s)							
MCS index	Modulation type	Coding rate	20 MHz channels		40 MHz channels		80 MHz channels		160 MHz channels	
			800 ns GI	400 ns GI	800 ns GI	400 ns GI	800 ns GI	400 ns GI	800 ns GI	400 ns GI
0	BPSK	1/2	6.5	7.2	13.5	15	29.3	32.5	58.5	65
1	QPSK	1/2	13	14.4	27	30	58.5	65	117	130
2	QPSK	3/4	19.5	21.7	40.5	45	87.8	97.5	175.5	195
3	16-QAM	1/2	26	28.9	54	60	117	130	234	260
4	16-QAM	3/4	39	43.3	81	90	175.5	195	351	390
5	64-QAM	2/3	52	57.8	108	120	234	260	468	520
6	64-QAM	3/4	58.5	65	121.5	135	263.3	292.5	526.5	585
7	64-QAM	5/6	65	72.2	135	150	292.5	325	585	650
8	256-QAM	3/4	78	86.7	162	180	351	390	702	780
9	256-QAM	5/6	N/A	N/A	180	200	390	433.3	780	866.7

Source Wikipedia

The table above gives a summary of the bit rates of different forms of QAM and PSK.

QAM Noise Margin
Higher order modulation rates are able to offer much faster data rates and higher levels of spectral efficiency for the radio communications system. However, they are considerably less resilient to noise and interference.

The ability of a Wi-Fi system to perform rapid "rate adaptation" (to adapt the modulation scheme to obtain the highest data rate for the given conditions) is an important area of evaluation. As signal to noise ratios decrease, errors will increase along with re-sends of the data, slowing throughput. By intelligently reverting to a lower order modulation scheme, the link can be made more reliable with fewer data errors and re-sends. Some system designers tune a Wi-Fi environment by eliminating higher data rates in extremely congested environments to minimize rate adaptation and the related retries. It is important to have this tool in your toolkit.

When Does This Technology Get Here

For most of us, the adoption of wireless technology was modest until 2003, when network speeds jumped up to 54 megabits per second with the availability of products supporting the new of the IEEE 802.11g communication standard. The technology for mobility was present, the need for mobility was evolving and the industry was ready for the adoption explosion to start.

During this time, we may have started deploying mobility-enhanced applications such as mobile ticket scanning and some level of sales application. If wireless was deployed at a venue, it was most likely in a limited way, either in both the Press Box and corporate office conference rooms or only in one of the two.

While still a hot amenity in most venues, wireless deployments moved quickly into industries that benefited from the increased productivity mobility provided. Adoption of pervasive Wi-Fi coverage by the young and innovative, especially those in college and K-12 environments, and the rapid acceptance and use of social media applications, started to change the fans' expectations. These generations have been raised on the expectation that wireless will be available everywhere, and they are your newest and most network active, innovative fans.

For the fans of the future Wi-Fi will not just be an expectation is will be considered and entitlement!

Enterprise Access Points Unit Shipments

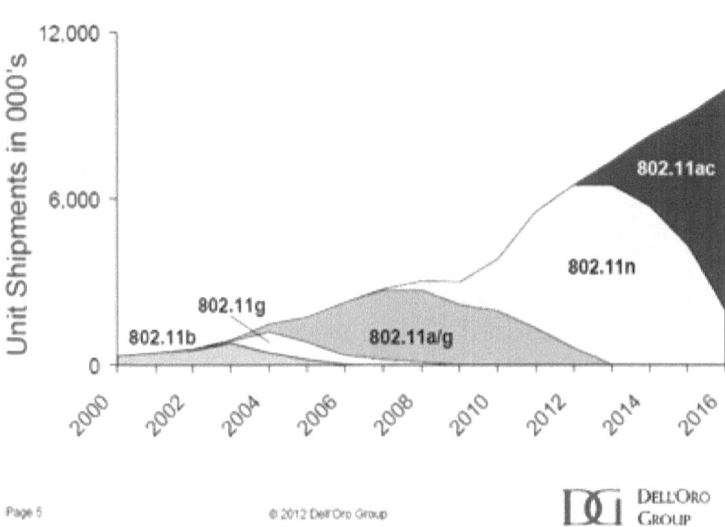

© 2012 Dell'Oro Group

DELL'ORO
GROUP

Source via Meru Networks

As of the second half of 2013, the first 802.11ac products are now reaching customers. Early reviews show peak over the air Internet access speeds of over 400 mbps. Based on the IEEE ratification schedule, it appears we are observing a nine-month delay between standards ratification and enterprise grade product availability.

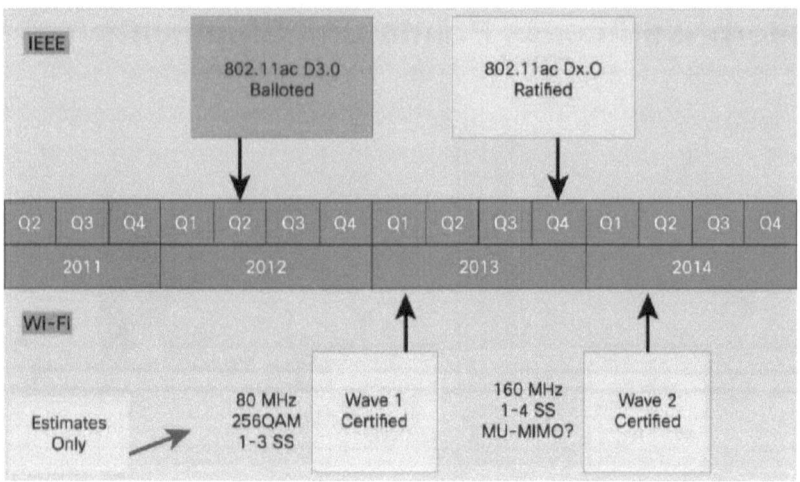

Standards Committee Time Lines   **Source IEEE web site**

If you are expanding your environment to accommodate in-season upgrades, the 802.11n technology will be well suited for the current equipment your fans will be using and their current expectations of Wi-Fi performance. If you are planning an upgrade for the off-season or next year's deployment, the first generation of 802.11ac products should provide a wonderful platform for enhancing the needs of the business while improving the fan experience. Any longer term project should seriously consider the 802.11ac generation two products.

Wave Two – The other half of the 802.11ac – Story

The rate of change continues. In the next wave of commercial quality access points there will be more radios, more special steams plus two key features will be added to the next wave 802.11ac product. Beam Forming and Multi-User MIMO. Of the two MU, MIMO will have the largest impact in the high density environment. This will be discussed in detail online and expanded upon in this book's second edition.

For Bonus Material on 802.11ac wave two visit
www.supportthebusiunessdelightthefans.com

HT MCS Index	Modulation and Coding Rate	Spatial Streams	Data Rate (MBPS)								VHT MCS Index
			20 MHz Chan		40 MHz Chan		80 MHz Chan		160 MHz Chan		
			No SGI	SGI	No SGI	SGI	No SGI	SGI	No SGI	SGI	
0	BPSK 1/2	1	6.5	7.2	13.5	15.0	29.3	32.5	58.5	65.0	0
1	QPSK 1/2	1	13.0	14.4	27.0	30.0	58.5	65.0	117.0	130.0	1
2	QPSK 3/4	1	19.5	21.7	40.5	45.0	87.8	97.5	175.5	195.0	2
3	16-QAM 1/2	1	26.0	28.9	54.0	60.0	117.0	130.0	234.0	260.0	3
4	16-QAM 3/4	1	39.0	43.3	81.0	90.0	175.5	195.0	351.0	390.0	4
5	64-QAM 2/3	1	52.0	57.8	108.0	120.0	234.0	260.0	468.0	520.0	5
6	64-QAM 3/4	1	58.5	65.0	121.5	135.0	263.3	292.5	526.5	585.0	6
7	64-QAM 5/6	1	65.0	72.2	135.0	150.0	292.5	325.0	585.0	650.0	7
	256-QAM 3/4	1	78.0	86.7	162.0	180.0	351.0	390.0	702.0	780.0	8
	256-QAM 5/6	1	n/v	n/v	180.0	200.0	390.0	433.3	780.0	866.7	9
8	BPSK 1/2	2	13.0	14.4	27.0	30.0	58.5	65.0	117.0	130.0	0
9	QPSK 1/2	2	26.0	28.9	54.0	60.0	117.0	130.0	234.0	260.0	1
10	QPSK 3/4	2	39.0	43.3	81.0	90.0	175.5	195.0	351.0	390.0	2
11	16-QAM 1/2	2	52.0	57.8	108.0	120.0	234.0	260.0	468.0	520.0	3
12	16-QAM 3/4	2	78.0	86.7	162.0	180.0	351.0	390.0	702.0	780.0	4
13	64-QAM 2/3	2	104.0	115.6	216.0	240.0	468.0	520.0	936.0	1040.0	5
14	64-QAM 3/4	2	117.0	130.0	243.0	270.0	526.5	585.0	1053.0	1170.0	6
15	64-QAM 5/6	2	130.0	144.4	270.0	300.0	585.0	650.0	1170.0	1300.0	7
	256-QAM 3/4	2	156.0	173.3	324.0	360.0	702.0	780.0	1404.0	1560.0	8
	256-QAM 5/6	2	n/v	n/v	360.0	400.0	780.0	866.7	1560.0	1733.3	9
16	BPSK 1/2	3	19.5	21.7	40.5	45.0	87.8	97.5	175.5	195.0	0
17	QPSK 1/2	3	39.0	43.3	81.0	90.0	175.5	195.0	351.0	390.0	1
18	QPSK 3/4	3	58.5	65.0	121.5	135.0	263.3	292.5	526.5	585.0	2
19	16-QAM 1/2	3	78.0	86.7	162.0	180.0	351.0	390.0	702.0	780.0	3
20	16-QAM 3/4	3	117.0	130.0	243.0	270.0	526.5	585.0	1053.0	1170.0	4
21	64-QAM 2/3	3	156.0	173.3	324.0	360.0	702.0	780.0	1404.0	1560.0	5
22	64-QAM 3/4	3	175.5	195.0	364.5	405.0	n/v	n/v	1579.5	1755.0	6
23	64-QAM 5/6	3	195.0	216.7	405.0	450.0	877.5	975.0	1755.0	1950.0	7
	256-QAM 3/4	3	234.0	260.0	486.0	540.0	1053.0	1170.0	2106.0	2340.0	8
	256-QAM 5/6	3	260.0	288.9	540.0	600.0	1170.0	1300.0	n/v	n/v	9

IEEE 802.11n MCS table - WITS via Aerohive Blog

Talk about anything you want – then-

Support the Business Delight the Fans

Nunca Mural – South beach Florida

Summary

No technology has moved faster from innovation to faster than wireless networking. Fans believe they are entitled to it. The business is hard pressed to run without it. There is an immediate need for a pervasive high capacity venue Wi-Fi network in Sports and Entertainment venues.

Wi-Fi is just as needed to support the front of the house activities directly support the players and enhance the fan experience, as it s to support the back of house running of the venue. Support the business first to generate in venue revenue then Delight the fans with an in venue multi-sensory communal experience they can share with friends.

Wi-Fi users must receive a strong, clear, actionable signal no matter how demanding the network's usage load, so Design for signal strength Symmetry - you cannot attain perfect symmetry you can only strive for it.

Deploy the fastest highest capacity network your budget can afford like the current state of the art is 802.11ac. Any deployment project planned for a year or more out from the now should seriously consider the 802.11ac wave two products.

This is not easy. The technology and business requirements are always changing. Wireless network in venues serve two very demanding masters: the business and the fans. The wireless is always blamed when even the end user cannot get service, regardless of the cause. So, plan for network management tools and Game Day wireless support services.

You can design the network by logical function, regularly update it with the best technology, regularly tune out the bad and tune in the good, fully monitor and support the network with good people then you can wake up each game day and say "Good morning, Lord!" or you cannot put in the time, money and effort into the Wi-Fi network and wake up each game day and say "Good Lord, it's a morning."

Enjoy

Main Sources

802.11 Fundamentals by Roshan, Leary, 802.11 Definitive Guide byM. Gast, CWNA Study Guide, Sports Fan 2.0 by Sutera, Social Media in Sports Marketing by Newman, Peck, Harris & Wihide, The Birth of ESPN by Bill Rasmussen, 13 Ways to Accomplish More by Doing Less byWilliam Teh, The wiki, Bharghavan Vadavur YouTube videos, Manufacturer web sites including Cisco, Aruba, Meru, Motorola, Ruckus, Enterasys, Aerohive, Ericson, ARS technical, Facebook, Fan Cost Index.

Great Wi-Fi in Stadiums is very satisfying when you get it right. It is a continuous effort of testing, refinement, and focus.

Mark has inspired me by his effort in getting the best coaches, continuously improving, and staying in the game until you win the Gold.

Mark Anthony McKoy – Olympic gold medalist - 110 meters hurdles at the 1992 Summer Olympics.

www.ingramcontent.com/pod-product-compliance
Lightning Source LLC
Chambersburg PA
CBHW030856180526
45163CB00004B/1605